Measurement in Medicine

Measurement in Medicine

Philosophical Essays on Assessment and Evaluation

Edited by
Leah McClimans

ROWMAN &
LITTLEFIELD
INTERNATIONAL

London • New York

Published by Rowman & Littlefield International, Ltd.
Unit A, Whitacre Mews, 26-34 Stannary Street, London SE11 4AB
www.rowmaninternational.com

Rowman & Littlefield International, Ltd. is an affiliate of Rowman & Littlefield
4501 Forbes Boulevard, Suite 200, Lanham, Maryland 20706, USA
With additional offices in Boulder, New York, Toronto (Canada), and Plymouth (UK)
www.rowman.com

British Library Cataloguing in Publication Data
A catalogue record for this book is available from the British Library

ISBN: HB 978-1-78348-847-6
PB 978-1-78348-848-3

Library of Congress Cataloging-in-Publication Data

Names: McClimans, Leah, editor.
Title: Measurement in medicine : philosophical essays on assessment and evaluation / edited by Leah McClimans.
Description: London ; New York : Rowman and Littlefield International, [2017] | Includes bibliographical and index.
Identifiers: LCCN 2017012208 (print) | LCCN 2017023973 (ebook) | ISBN 9781783488490 (Electronic) | ISBN 9781783488476 (cloth : alk. paper) | ISBN 9781783488483 (pbk. : alk. paper)
Subjects: LCSH: Medical care--Evaluation--Methodology. | Clinical medicine--Statistical methods.
Classification: LCC RA399.A1 (ebook) | LCC RA399.A1 M424 2017 (print) | DDC 610.72--dc23
LC record available at https://lccn.loc.gov/2017012208

♾™ The paper used in this publication meets the minimum requirements of American National Standard for Information Sciences—Permanence of Paper for Printed Library Materials, ANSI/NISO Z39.48-1992.

Printed in the United States of America

Contents

Introduction

Leah M. McClimans

In 1992, with the publication of "Evidence-Based Medicine: A New Approach to Teaching the Practice of Medicine," The Evidence-Based Medicine Working Group and the *Journal of the American Medical Association* famously ushered in a "paradigm shift." This shift emphasized evidence over experience, statistics over intuition, and method over judgment. It is interesting that although this new paradigm prioritized quantitative evidence, the role of measurement does not appear in its early formulation. Yet we might understand this paradigm shift to evidence-based medicine (EBM) as a shift toward a reliance on measurement rather than a simple shift toward evidence. After all, evidence need not be quantitative, but the EBM movement emphasizes measurable, quantitative evidence. (McClimans 2013, 521).

Medicine is not alone in making this move to favor measureable outcomes. Indeed, the maturation of the sciences is often associated with such a move. In the late eighteenth century, the physical sciences were "quantified," and contemporary experimental physics is now largely synonymous with mathematical outcomes (Kuhn 1977, 213–21). It is perhaps not surprising that chemical, biological, social, and medical sciences have sought to move in this same direction, a direction that relies increasingly on precision and measurement (Wise 1995).

Measurement raises several epistemological, metaphysical, ethical, and social questions, questions that intersect with multiple subfields in philosophy. Traditionally, philosophers of the natural sciences have turned their attention to epistemological and metaphysical issues of measurement (e.g., Chang 2004; Tal 2013; Van Fraassen 2010), whereas philosophers of economics politics, and ethics have turned to measurement issues that raise largely normative questions (e.g., Broome 2004; Sen 2001; Sumner 1996). Measurement in medicine complicates this division of labor.

Much has been written about whether medicine is an "art" or a "science," and although the rise of EBM and measurement would seem to settle the question in favor of the latter (for an alternative point of view, see Montgomery 2005), it remains the case that in medicine, people are at the center of all that is done, including measurement. Consider some of the different ways in which measures figure in medicine: clinicians measure patients' heart rate and blood pressure to provide indications of individual health; epidemiologists measure the distribution of disease in a population to better understand its cause; health service researchers measure health outcomes of people to understand better questions of quality for a population; and health economists measure the quantity and quality of lives to allocate health care resources in a cost-effective manner. These examples illustrate how patients are both the subject of medical measurements and their purpose (e.g., we measure the health outcomes of women who have undergone reconstruction following mastectomy to better serve future women in a similar situation). But because people are the subject and purpose of medical measurement, epistemic and metaphysical questions about how and what is measured cannot do justice to all the relevant measurement considerations. Similarly, because medical outcomes, such as morbidity and quality of life, involve natural processes (e.g., physiological and psychological), ethical and social questions do not cover all that is relevant.

If we want to understand and critically evaluate the EBM movement, we need to tackle measurement, but if we want to tackle measurement in the EBM movement, then we need to mobilize the full spectrum that philosophy offers (i.e., ethical, social, epistemic, and metaphysical fields of study). This volume aims to bring together essays that offer such a range of philosophical work. Thus, measurement, the achievement of the quantitative sciences, becomes a meeting point where topics from the philosophy of medicine (i.e., metaphysics and epistemology) meet those traditionally found in medical ethics and value-based practice (see Fulford 2008). Indeed, Smart's contribution in the beginning of this volume sets the stage in this regard providing one perspective on how measurement and the hierarchy of evidence intersect with values and theory.

The essays in this book are gathered together into three sections with each one illustrating a distinct, but overlapping, area of contemporary research within the philosophy of medical measurement broadly understood: measurement and evidence-based medicine, measuring instruments, and measurement and policy.

MEASUREMENT AND EVIDENCE-BASED MEDICINE

When philosophers of science first started to consider topics in medicine, they brought to bear the interests that sustained them in the natural sciences (e.g., interests in causation, explanation, modeling). Initially, these areas of interest led philosophers of medicine to focus significantly on questions regarding evidence and study design (e.g., hierarchies of evidence, the methodology of clinical trials). Recognition that measurement held philosophical as opposed to merely practical value in these considerations was slow to be acknowledged. When it was acknowledged, it was mainly through the pioneering work of Alex Broadbent and his introduction of a new subspecialty: philosophy of epidemiology (Broadbent 2013).

Epidemiology studies the distribution and determinants of health outcomes in human populations. In his work, Broadbent (e.g., 2013; this volume, Chapter 2) emphasizes the important role that causation plays in establishing the determinants of disease. Epidemiologists use analytic measures to express the strength of association between two variables (i.e., the strength of association between exposure and outcome). By comparing different groups of people, epidemiologists ask whether differences in health outcomes are the result of differences in exposure (e.g., in comparing children who are breastfed with children who are not breastfed, epidemiologists ask whether this difference in infant feeding is associated with certain outcomes, for instance, diarrheal diseases or asthma [e.g., Dogaru et al. 2014; Lamberti et al. 2011]).

There are several analytic measures that can be used to express this association (e.g., relative risk, odds ratio, population excess fraction). However, sometimes these measures are not simply used to describe a strength of association between exposure and outcome; sometimes they are used to quantify the strength of the causal relationship between them (e.g., breastfeeding protects children from asthma). Thus, the philosophical questions are when do these analytic measures quantify the strength of a causal relationship, and when do they simply represent the association between two variables? And what do measures that quantify causal strength measure?

As Broadbent (2013) explains, the mathematics of the analytic measures do not answer these questions. For instance, say we calculate the relative risk of asthma among formula-fed infants such that ten times as many bottle-fed infants develop asthma during childhood than breastfed infants. This calculation tells us nothing about the causal relationship between bottle feeding and asthma, yet the point of the calculation is to provide causal evidence to, for example, support public health policies that promote breastfeeding. As Broadbent (2013, 31) puts it, there are causal assumptions implied by using "exposure" and "outcome" that go beyond the math and require philosophical analysis.

Broadbent's work introduced philosophers to analytic measures of association and showed why they warranted philosophical interest. Central to his work are questions of causal inference. Broadbent's contribution to this volume as well as Stegenga and Kenna's, and Mncube's build in different ways on the question of causal inference in the context of statistical measures. They illustrate one rich area of research that focuses on the evidentiary claims made across different areas of medical research and study design.

MEASURING INSTRUMENTS

Epidemiologists use analytic measures to express the strength of association between exposure and outcome, but what are outcomes and how are they measured? Earlier I gave the example of diarrheal diseases and asthma as two specific outcomes. More generally, the primary outcomes of interest in medicine and surgery are mortality, morbidity, and quality of life. These outcomes are themselves measured using a variety of instruments (e.g., quality of life is typically measured with self-reported questionnaires, morbidity may be inferred from a sphygmomanometer reading, physician questionnaire, or diagnostic test). These measuring instruments are used within a clinical study to provide quantitative information about a particular health outcome. Within the context of a specific study design the results from these instruments are reported using a statistic (e.g., relative risk, difference between means, Quality Adjusted Life Year [QALY]). Measuring instruments, as I am using them, thus provide some of the empirical data for further measurement (e.g., analytic measures, summary measures such as the QALY, and so on).

The notion of a "measuring instrument" covers a wide range of technology from weighing scales to assays to questionnaires. The "objects" that these instruments aim to measure also vary widely from, for example, mass to sex to quality of life. One useful distinction among these "objects" of measurement is between pinpoint and *Ballung* concepts (Bradburn, Cartwright, and Fuller, this volume, Chapter 5). Pinpoint concepts, such as mass, sex, and mortality, are quantitative or qualitative features, the aim of which is to define them precisely so that individuals fall into discreet categories that can be compared across populations and time. *Ballung* concepts, such as quality of life, are difficult to define precisely, often have normative implications, and function differently in different contexts, thus making them a challenge to generalize across populations and time.

The development and application of measuring instruments for *Ballung* concepts is a good example of how philosophical analysis of measurement in medicine acts as a bridge between the concerns of epistemology and metaphysics and those of ethical and social/political philosophy. When we evaluate quality of life measures, for instance, we may question their validity—an epistemic concern—but we might do so based on the measures' ability or inability, for instance, to incorporate response shift (Barclay-Goddard and Epstein 2009) (i.e., patients' changing understanding of the target construct over the course of a disease or treatment into the measurement outcome). This concern about response shift can stem from ethical questions regarding the instrument's ability to fulfill its purpose in reporting the patients' point of view. Additionally, we can question the ontology of quality of life and wonder whether such a concept can be measured at all. If it cannot be measured, then social and resource distribution questions are raised regarding how to incorporate quality of life (and not simply quantity of life) into clinical and cost-effectiveness calculations, drug labeling claims, health care resource management, and so on.

The measuring instruments used in medicine are a fertile area of research, which bridge different philosophical concerns. In this section, contributions by McClimans, Cupples, Engebretsen and Heggen, and Upshur focus on the measurement of different *Ballung* concepts. These contributions are representative of a growing area in the philosophy of medicine that question the methodology of these instruments while recognizing the need medicine has for such information.

MEASUREMENT AND POLICY

The connection between health measurement and health policy probably requires the shortest introduction to a philosophical audience. Traditionally, philosophers who studied measurement in medicine and health care have focused on population health and priority setting. There is a vast and excellent philosophical literature concerned with these issues. John Broome's *Weighing Lives* (2004), Daniel Hausmans's *Valuing Health* (2015), and John Harris's much-cited "QALYfying the value of life" (1987) are just three examples of this extensive literature.

One characteristic of this work, which differs from the work on measures of causal strength or even measuring instruments of *Ballung* concepts, is the persistent normative dimension of the questions that sustain and populate it. What should be the object of measurement when considering population health? How should we measure population health? Should we solicit information on preferences or subjective experience or capabilities? When considering summary measures of population health, should we focus on the years lost due to poor quality of life as we do with Disability Adjusted Life Years (DALYs), or should we focus on years gained due to improved quality of life as we do with QALYs? When making resource allocation decisions, should we use cost-effectiveness analysis? And if we do, what are the moral objections to such an analysis, and how can they be overcome?

The practical upshot of the normativity inherent in these questions is that philosophical treatment of measurement in health policy is more oriented toward developing and applying metrics that reflect our sense of what is right, good, and/or just. This is different from the standard treatment of pinpoint or *Ballung* measuring instruments, which tend to be conceptualized as tethered to natural properties or processes (i.e., physiological, psychological). Yet certain measuring instruments play a significant role in the measurement of health policy, thus bridging these differences. Just as some outcome measures provide the empirical data for analytic measures, utility measures of quality of life, such as the EQ-5D (EuroQoL), provide empirical data for QALYs.

Utility measures are always index measures; this means that the sum of an instrument's indications can be added together to produce a single number. In the context of a utility measure, the index score is correlated with utility values that are derived from the public's ranking of different health states to yield a measurement outcome. The public ranks health states using thought experiments, such as the time trade-off technique. The EQ-5D (EuroQol) is an example of a utility measure of quality of life that uses the time trade-off technique (Brooks 1996).

QALYs are a way of valuing and comparing health outcomes. They combine the length of survival with a measure of the quality of that survival and assume that given a choice, a person would choose a shorter life of high quality to a longer life of poor quality. A year of life in perfect health is associated with 1 QALY, and death is associated with 0. To determine the values in between, a measuring instrument, such as the EQ-5D (EuroQol), provides the indications that are correlated with the utility values derived from the public's ranking. These respective utility values are multiplied by the number of years that patients are expected to live in a particular state of health. This multiplication provides a QALY between 1 and 0. QALYs can then be used to compare the quality of life of different individuals in a population or the same individual given different treatment regimes. To determine the cost/QALY, or cost effectiveness, we divide the difference of the cost of the interventions or treatments in question by the difference of their QALY gains.

Measurement and health policy is a well-developed area of philosophical research. Badano, John, and Junghans's, and Hausmans's contributions to this section explore normative and practical questions regarding the QALY and cost effectiveness. Dickson's contribution provides a layer of complexity to the QALY in his consideration of the Value of a Statistical Life (VSL). In Plutynski's contribution, she contextualizes health policy not in terms of priority setting, but rather in terms of the quality of care provided in hospital settings. She challenges our current metrics for determining when care is effective by arguing that we need to move beyond the assessment of individual, consecutive outcomes and begin thinking about effectiveness in terms of more holistic processes.

FUTURE DIRECTIONS

This volume brings together for the first time emerging and established areas of philosophical work in medical measurement. These contributions represent diverse interests from researchers with different academic backgrounds, but together they tell a story about how measurement functions at different levels within medicine and health policy. This story also highlights different ways measures can be conceptualized (e.g., as measuring the strength of a causal relationship, a property or process, or a summary of quantity and quality of life). Finally, it underscores the different kind of concerns that preoccupy different measurement practices (e.g., epistemological, metaphysical, ethical, and social).

But what of the future for measurement in medicine? In addition to the work presented here, what areas of medicine and health policy have been overlooked by philosophers interested in measurement? I think there are at least three such places. First, although philosophers of science have devoted considerable attention to the epistemic ramifications of technological artifacts, such as thermometers and clocks (e.g., Chang 2004; Tal 2014), philosophers of medicine have not given the same attention to medical technologies such as sphygmomanometers or magnetic resonance imaging (MRI). If the philosophical fruits to come out of the study of thermometers and clocks are any indication of what we can learn from an investigation of such technologies, then we ought to pursue such work in the medical context.

Second, when philosophers and the philosophically inclined discuss health policy, they tend to focus on priority setting and cost effectiveness. But as Plutynski's contribution to this volume indicates, there is more to be said about health measurement and policy than considerations of resource allocation. Health service research (HSR) plays an important, but philosophically neglected, role in shaping health policy. Health service researchers study the effectiveness of different organizations of health care (e.g., how many emergency rooms are required to serve a particular state or county?). They ask how people gain access to health care and are interested in what happens to them when they do (i.e., what is the quality and continuity of health care in different institutions?). HSR represents a large and influential sector within health care that utilizes a diverse array of measures to understand health delivery, access, and organization to shape policy. Philosophy has all but ignored it.

Third, some (see Hobart et al. 2007) have argued that the failure of clinical trials to yield larger numbers of effective treatments may be due to the lack of scientific rigor of their measuring instruments. It should not be surprising that pharmaceutical companies keen to demonstrate the effectiveness of their products while using measures that will satisfy FDA guidelines are eager to explore methods that will improve their success. And it is not only industry that wants to see the acceleration of medical product development; the FDA also shares this goal.

Public-private partnerships, such as Critical Path Institute (C-Path), created under the auspices of the FDA's critical path initiative program, aim to create drug development tools (DDT) (e.g., new data, measures, and methodological standards) to accelerate the pace and reduce the cost of medical product development (Critical Path Institute 2015). Philosophers interested in measurement ought to be interested in, for example, how these kinds of partnerships shape the FDA's guidelines for using measurement outcomes in drug labeling claims. Measurement and industry is thus the third area of research that philosophers interested in measurement should consider.

In his contribution to this volume, Upshur (Chapter 9) recalls a common exhortation at meetings with clinicians and policy leaders: "you cannot manage what you cannot measure." Even if this claim is not literally true, it is increasingly a practical fact: medical management demands measurement. For those interested in pursuing philosophical questions of measurement practices, this ubiquity is good news. Indeed, many aspects of medicine are not obviously or easily measured. The pressure to quantify these areas of medicine will require ingenuity. Thus, they will not only expand our understanding of, for example, the epistemology of measurement, but also open up opportunities for philosophers to offer practical and conceptual support to these endeavors.

REFERENCES

Barclay-Goddard, Ruth, and Joshua D. Epstein. 2009. "Response Shift." *Quality of Life Research* 18: 335–46.

Broadbent, Alex. 2013. *Philosophy of Epidemiology*. London: Palgrave Macmillan.

Brooks, Richard. 1996. "EuroQol: The Current State of Play." *Health Policy* 37: 53–72.

Broome, John. 2004. *Weighing Lives*. Oxford: Oxford University Press.

Chang, Hasok. 2004. *Inventing Temperature: Measurement and Scientific Progress*. Oxford: Oxford University Press.

Critical Path Institute. 2015. "About Critical Path Institute." Accessed April 7, 2016. http://c-path.org/about/.

Dogaru, Cristian M., Denise Nyffenegger, Anina M. Pescatore, Ben D. Spycher, and Claudia E. Kuehni. 2014. "Breastfeeding and Childhood Asthma: Systematic Review and Meta-Analysis." *American Journal of Epidemiology* 179: 1153–67.

Fulford, Kenneth W. M. 2008. "Values-Based Practice: A New Partner to Evidence-Based Practice and a First for Psychiatry?" *Mens Sana Monographs* 6: 10–21. doi:10.4103/0973-1229.40565.

Harris, John. 1987. "QALYfying the Value of Life." *Journal of Medical Ethics* 13: 117–23.
Hausman, Daniel M. 2015. *Valuing Health: Well-Being, Freedom, and Suffering*. Oxford: Oxford University Press.
Hobart, Jeremy C., Stefan J. Cano, John P. Zajicek, and Alan J. Thompson. 2007. "Rating Scales as Outcome Measures for Clinical Trials in Neurology: Problems, Solutions, and Recommendations." *Lancet Neurology* 6: 1094–105. doi:10.1016/S1474-4422(07)70290-9.
Kuhn, Thomas. 1977. *The Essential Tension: Selected Studies in Scientific Tradition and Change*. Chicago: Chicago University Press.
Lamberti, Laura M., Christa L. Fischer Walker, Adi Noiman, Cesar Victora, and Robert E. Black. 2011. "Breastfeeding and the Risk for Diarrhea Morbidity and Mortality." *BMC Public Health* 11: S15. doi: 10.1186/1471-2458-11-S3–S15.
McClimans, Leah. 2013. "The Role of Measurement in Establishing Evidence." *Journal of Medicine and Philosophy* 38: 520–38. doi:10.1093/jmp/jht041.
Montgomery, Kathryn. 2005. *How Doctors Think: Clinical Judgment and the Practice of Medicine*. New York: Oxford University Press.
Sen, Amartya. 2001. *Development as Freedom*. Oxford: Oxford University Press.
Sumner, Leonard. 1996. *Welfare, Happiness and Ethics*. Oxford: Clarendon Press.
Tal, Eran. 2013. "Old and New Problems in Philosophy of Measurement." *Philosophy Compass* 8: 1159–73. doi:10.1111/phc3.12089.
Tal, Eran. 2014. "Making Time: A Study in the Epistemology of Measurement." *The British Journal for the Philosophy of Science* 67: 297–335.
Van Fraassen, Bas C. 2010. *Scientific Representation: Paradoxes of Perspective*. Oxford: Clarendon Press.
Wise, M. Norton. 1995. *The Values of Precision Exactitude*. Princeton, NJ: Princeton University Press.

Part I

MEASUREMENT AND EVIDENCE-BASED MEDICINE

Chapter 1

How Evidence-Based Medicine Highlights Connections between Measurement and Evidence

Benjamin Smart

Evidence-based medicine (EBM) holds that the evidence provided by epidemiological studies provides better justification for clinical decisions than expert opinion, intuition, or pathological reasoning. It operates on the premise that whether an individual should be treated in a particular way depends primarily upon whether clinical trials have shown the treatment to be beneficial, effective, and cost efficient relative to the alternatives (Sackett et al. 1996). Evidence-based practitioners (those who endorse and practice EBM) thus argue that EBM limits the risk of prescribing inefficient and/or harmful treatments and thus improves mortality and morbidity rates as well as quality of life.

EBM is often associated with a hierarchy of evidence. At the top of the hierarchy sit randomized controlled trials (RCTs), followed by observational studies, such as case-control studies and cohort studies. At the bottom of the evidence hierarchy, one finds mechanistic and pathophysiological reasoning, clinical judgment, and expert opinion/intuition (Cartwright 2007; Howick 2011). Although all respectable forms of medicine are, in some sense, evidence based, the best evidence, as prioritized by the EBM practitioner, is that provided by clinical epidemiologists. Their conclusions are drawn from data collected from carefully designed studies, which, of course, one can obtain only through "measurement"; it is unsurprising, then, that Leah McClimans (2013, 521) has suggested that the paradigm shift to EBM is not so much a shift toward the reliance on evidence as it is a shift toward reliance on measurement.

In this chapter, I highlight how decision making in clinical medicine is vulnerable to implicit bias, but argue that the paradigm shift toward measurement discussed by McClimans goes some way to healing this serious problem in health care. I examine the evidence-based practitioner's approach to evidence in medicine, and in exploring the theory laden nature of both evidence and measurement, I highlight the connection between evidence and measurement that EBM brings into focus.

EVIDENCE-BASED MEDICINE, PATIENT-CENTERED CARE, AND IMPLICIT BIAS

Evidence-Based Medicine

The term "evidence-based medicine" first appeared in a paper by Gordan Guyatt in 1991, but basing clinical decisions on evidence is hardly a new strategy in Western medicine.[1] Indeed, for those skeptical of certain forms of alternative medicine, basing clinical decisions (whether diagnostic, prognostic, or treatment-related) on good evidence is arguably that which distinguishes proper medical practice from that of witch doctors and swindlers.

Crudely speaking, EBM can be viewed as the explicit and judicious use of "best evidence" in a clinical setting (e.g., Sackett 1996). Much, but not all, of this "best" evidence is gathered through measurements in large-scale clinical trials. The evidence obtained translates into best practices and clinical guidelines, which are then implemented by all evidence-based practitioners to try to standardize clinical decisions. This approach is designed to minimize the role of intuition and guesswork in medical practice, which practitioners of EBM assume to be harmful. As I discuss later in this chapter, other forms of evidence are admissible in evidence-based practice, but EBM ultimately dictates that wherever possible, clinical decisions concerning interventions (or lack thereof) should primarily be made in accordance with the clinical guidelines formed in light of RCTs and observational studies, that is, in accordance with the measurements made and interpreted by epidemiologists.

One should note that this is not to say that clinical expertise is irrelevant to evidence-based practice, as some critics of EBM imply (e.g., Karthikeyan and Pais 2010). Good doctors, writes David Sackett and colleagues (1996, 71),

> use both individual clinical expertise and the best available external evidence, and neither alone is enough. Without clinical expertise, practice risks becoming tyrannized by evidence, for even excellent external evidence may be inapplicable to or inappropriate for an individual patient. Without current best evidence, practice risks becoming rapidly out of date, to the detriment of patients.

EBM thus requires clinicians to use their clinical expertise to listen to patients, to make the right measurements, and to identify the correct best practices and guidelines to follow.

The alternative paradigm to EBM in modern medicine is "patient-centered care." Clinical decision making, for advocates of patient-centered care, prioritizes the values and needs of the patient and involves more patient participation in treatment decisions. This differs from EBM in that whereas EBM's primary concern is disease (the causes of disease, preventive and curative interventions, and so on), patient-centered care has "a humanistic, biopsychosocial perspective, combining ethical values on the 'ideal physician,' with psychotherapeutic theories on facilitating patients' disclosure of real worries, and negotiation theories on decision making" (Bensing, 2000, 17).

Just as evidence-based practitioners are not entirely robotic in their approach to health care (EBM also respects patients' values; e.g., if a patient is particularly averse to the treatment clinical trials suggest is best [for religious reasons, for example], then evidence-based practitioners will certainly not force that treatment on their patient), those who favor patient-centered care do not altogether ignore biomedical evidence. To ignore *all* evidence, in this way, would be to rely entirely on luck for both diagnosis and treatment. Nevertheless, EBM's heavier reliance on the evidence produced by clinical trials is indicative of its heavier reliance on measurement.

Implicit Bias and Medicine

Implicit biases, crudely speaking, are unconscious prejudices that affect behavior such that one (unconsciously) discriminates against members of a socially stigmatized group (Brownstein 2015). Although an individual may (and as far as they're concerned, honestly) claim not to be prejudiced against a particular ethnic group, socioeconomic group, or sex, examination of that individual's behavior often indicates otherwise, since social attitudes and stereotypes naturally embed unconscious biases.

Unfortunately, studies have convincingly shown that implicit biases play a significant role in clinical decisions. Green et al. (2007), for example, conducted a study in the United States comparing the thrombolysis decisions for black patients and white patients with acute coronary syndromes. The physicians showed a significant anti-black implicit bias, and further, although their *explicit* biases did not affect their decisions to treat patients, "as the degree of anti-black bias on the race preference [computer-based implicit association test] increased, recommendations for thrombolysis for black patients decreased" (Green et al. 2007, 1235). Similarly, Schulman et al. (1999) showed that a white female in the United States is far more likely to be recommended for cardiac catheterization by her physician than a black female with the same symptoms. In the United States, at least, the quality of health care received by ethnic minorities thus seems to be significantly lower than that received by their white compatriots (Stepanikova 2012). On the face of it, this can be put down primarily to the implicit biases of clinicians.[2]

In short, despite clinicians' explicit biases tending not to play a role in their clinical decision making, the studies above provide good evidence that clinicians (like the rest of us) are not impervious to implicit bias and, worryingly, these biases are prone to manifest in the form of inconsistent and/or poor clinical decisions.

Implicit Bias and EBM

When a practitioner is inclined toward patient-centered care, her clinical decisions are prima facie particularly vulnerable to implicit bias. Unlike patient-centered care, however, EBM is explicitly designed to encourage (where appropriate) consistency in diagnosis, prognosis, and treatment. On the assumption that the necessary trials have taken place, physicians practicing EBM will (in principle) treat patients with the same symptoms in the same way; that is, assuming (i) that the intervention(s) necessary to treat the condition associated with a set of symptoms has/have been identified by clinical trials and (ii) that these trials have informed the clinical guidelines that evidence-based practitioners are bound to follow, all clinical decisions are standardized so that implicit biases play little or no role in clinical decision making.[3] In principle, then, if all clinicians practiced EBM, situations such as those identified by Schulman et al. (1999) would not arise.[4] On the face of it, at least, the paradigm shift toward measurement is a shift away from the impact of implicit bias in clinical medicine.

IMPLICIT BIAS AND THE THEORY-LADENNESS OF MEASUREMENT IN MEDICINE

Thomas Kuhn argued extensively that scientific observation is theory-laden. What one perceives, deems salient, and describes is at least partly determined by prior scientific theory (Kuhn 1996). To measure velocity of an object, for example, one must at the very least (i) be able to measure (using some device designed for the purpose) the time taken for that object to travel a measured distance; (ii) be able to measure distances accurately; and (iii) understand the theory describing the relationship between time, distance, and velocity. The theory-ladenness of observation, claims Kuhn, extends to all measurements in all sciences.

Good and Bad Theory-Ladenness

Theory-ladenness is often unproblematic. After all, good theories and new technologies permit one to measure outcomes one previously could not. Without magnetic resonance imaging (MRI), for example, much of the information relevant to many patients' cancer treatments would be unobservable (i.e., the development of the theory underlying MRI scanners, and the machine itself, have doubtless saved many lives). The problems concerning theory-ladenness arise when either the theories employed are bad theories or our instruments are not very good at doing what we ask of them. Intuition is a good guide to what counts as good theory and what counts as bad, but nevertheless, below I list some likely qualities of both.

Good Theory-Ladenness

- Good theory-ladenness is truth conducive.
- Good theory-ladenness enables and/or improves the accuracy of measurement outcomes.[5]
- Good theory-ladenness provides higher quality evidence.
- Good theory-ladenness is conducive to good clinical decision making.

Bad Theory-Ladenness

- Bad theory-ladenness is not truth conducive.
- If the theory underlying a measurement is bad, then those measurement outcomes are likely to be inaccurate.
- Evidence laden with bad theory is poor evidence.
- Bad theory-ladenness is conducive to bad or, at the very least, unjustified clinical decisions.

Given the results of studies such as those considered above, it seems that the stigma attached to groups such as African Americans results in "bad theory." This bad theory manifests in the form of implicit bias, which unconsciously, but systematically, feeds into what clinicians perceive and what is deemed salient to clinical decision making[6] and ultimately can lead to bad decisions concerning prognoses and treatment. As we shall see, EBM attempts to address the consequences of bad forms of theory-ladenness in medicine through additional (strategic) theory.

Kuhn's Three Forms of Theory-Ladenness in Measurement

Consider the following three features of theory-laden observation/measurement identified by Kuhn and their applicability to the medical sciences.

Perceptual Theory Loading

Using the example of combustion, Kuhn notes that when two scientists are working in different paradigms, they perceive different things. The theory of combustion endorsed by Joseph Priestley supposed that combustion and respiration were the release of a substance called phlogiston. Antoine Lavoisier, on the other hand, inspired a paradigm shift to the oxidation theory of combustion. When Priestley observed a substance burning, his perceptual experience conformed to his phlogiston theory; that is, he would (in his mind, at least) perceive the release of phlogiston. But Lavoisier did not share Priestley's perceptual experience. He perceived an oxidation process conforming to his own theory of combustion (e.g., Bogen 2013).

So much for combustion, but how might this apply to medicine? In a discussion concerning the theory-ladenness of perception in psychiatry, Rachel Cooper (2004, 15) discusses studies examining the effect of contextual information on emotion recognition:

> [In the studies examined by Fernandez-Dols and Carroll 1997], subjects are shown a face, for example a woman crying, and are given information regarding the context, for example they might be told that the woman has been given a present, and then they are asked to judge the emotion that the person is probably experiencing.

Fernandez-Dols and Carroll's 1997 meta-analysis showed inconsistent results in the trials they studied (i.e., some showed the contextual information to have a significant effect and others did not), so as Cooper points out, when it comes to contextual information of this form we must remain agnostic about the effects of perceptual theory loading. However, both Green et al.'s (2007) study of the effect of pictures depicting the skin color of a participant and Schulman et al.'s (1999) study of treatment differences between black and white women provide significant results. Given that *visual* stimuli, such as skin color, are present in all clinical consultations, it is reasonable to conclude that perceptual theory loading (minimally in the form of implicit bias) at least sometimes affects clinical decisions.

Semantical Theory Loading

Scientific observations are described using scientific language (e.g., combustion, temperature, mass, velocity, and so on). However, although scientists tend to use the same terms, Kuhn has argued that scientists working within different paradigms understand those terms in different ways. Priestly, when using the word "combustion," understood the term to mean the release of phlogiston. Lavoisier understood it to mean a process of oxidation.

Although I will not defend this claim rigorously, semantical theory loading looks unlikely to be prominent in Western medicine. Medical students in the twenty-first century (in the West) understand scientific terms in roughly the same way since they all operate within a single paradigm. On the other hand, Western medical practitioners, homeopaths, and African traditional healers might understand the scientific terms in different ways, since they operate within different medical paradigms. This chapter is concerned only with modern Western medicine, however, so I will leave this issue alone.

The Aristotelian paradigm, explains Kuhn (1996, 123), requires the experimenter to measure "the weight of the stone, the vertical height to which it had been raised, and the time required for it to reach rest" and "ignore radius, angular displacement, and time per swing." But both angular displacement and time per swing were salient in the Galilean paradigm since that paradigm viewed pendulum swings as constrained circular motions (Bogen 2013). Neither angular displacement nor time per swing is relevant to the Aristotelian paradigm since the Aristotelian views pendula motion as objects "falling under constraint toward the center of the Earth" (Kuhn 1996, 123). Consequently, those working in the Aristotelian paradigm and those working in the Galilean paradigm would measure entirely different quantities when investigating the same phenomenon.

Just as in the above example from physics, practitioners working in different paradigms (e.g., modern medicine, holistic medicine, and traditional healers) are very likely to find different qualities salient and thus make different measurements. Within Western medicine, one might assume there to be a consensus as to what the salient qualities of a patient are, but this is not the case. Those who encourage patient-centered care consider the values, needs, and desires of patients to be far more salient in clinical decision making than do evidence-based practitioners.

Strategic Theory and the Theory-Ladenness of Evidence in EBM

When any evidence type is used in clinical practice, observations and measurements must be made. After all, one cannot use one's intuition or clinical experience without first observing the patient and her symptoms. Measurements themselves are thus a constant form of evidence in medicine, so it is no surprise that the theory-ladenness that comes with observation always plays a role in clinical decision making. EBM, however, brings a further form of theory in evidence to light. What I call *strategic theory* is often present in evidence used in medicine (and elsewhere). It is the theory that does not *directly* come from observation or measurement. There are at least two ways in which strategic theory enters the kind of evidence produced by epidemiologists. The first comes prior to the measurement process and the second after the data have been collected.

The first form of strategic theory-ladenness in EBM is to be found in the study-design process. Before an epidemiologist begins collecting measurement outcomes, she must decide what kind of study is most appropriate and how it must be conducted given its goals, the environment in which the measurement process can take place, ethical considerations, and so on. The study-design process is a form of strategic theory in evidence, since it takes place prior to the observation process.

The second form of strategic theory-ladenness in EBM comes post-measurement. Statistical techniques are employed by epidemiologists to collate measurement outcomes from a study and organize them into more accessible forms of evidence. This evidence sharply indicates whether the measurements support a particular hypothesis, and if they do, the evidence quantifies the strength of this support. Examples in medicine include mortality rate, morbidity rate, odds ratio, relative risk, excess risk, and attributable risk, but there are many more. The statistical strategies employed by epidemiologists to arrive at these forms of evidence are employed after the measurement process and thus count as strategic theory. When a study is well designed, such things as mortality rates can provide excellent evidence for the effectiveness of an intervention.

The evidence provided by epidemiological studies can be laden with bad strategic theory as well as good, however; just as the quality of evidence can be tainted by bad theory in the measurement process, so too it can be tainted by bad theory in the strategic processes pre- and post-measurement. This can lead to a high probability of the study's findings being inaccurate. Consider the following passage from Williams and Seed (1993, 317), who analyzed fifty-eight dermatological clinical trials with negative conclusions and the probabilities of those trials missing an effective treatment:

> All but one of the 44 evaluable trials had a greater than 1 in 10 risk of missing a 25% relative treatment difference (median risk 81%), and 31 of the trials (70%) were so small that they had a greater than 1 in 10 risk of missing a 50% relative treatment difference (median risk 42%). The "negative" trial result was compatible (within 95% confidence limits) with a 25% beneficial relative treatment effect in 36 studies (82%), and a 50% treatment benefit in 22 studies (50%). Only one study used confidence intervals to describe the main findings, and only three studies (7%) mentioned the basis for sample size estimation at the outset of the study.

Williams and Seed's study indicates that bad theory-ladenness is not only possible in trial-based evidence, but it is also common. In this case, to reduce the risk of missing an effective intervention, most of the dermatological trials studied required more participants, but there are many other ways in which poor study design can result in poor evidence. Bad theory-ladenness aside, however, it is worth looking at an example of the evidence strategic theory can provide.

Suppose one wished to investigate the causal effect of exposure S (say, burned toast) on population P with respect to disease D (say, bowel cancer). Here, the case group will include only participants from P suffering from D, and the control group, only participants from P not suffering from D. After taking the appropriate measurements, one might discover that the odds of exposure to S given D (i.e., the number of participants exposed to S divided by the number not exposed to S in the case group) are higher than the odds of exposure to S given not D (i.e., the number of participants exposed to S divided by the number not exposed to S in the control group). This indicates an association between the exposure S and disease D. If the odds of S are significantly higher in the control group than in the case group, then this association may indicate that S is a preventive intervention against D; if the odds of S are lower in the control group, then this association may indicate that S is a cause of D (e.g., Rothman, Greenland, and Lash 2008).

Epidemiologists express the degree of association in terms of the odds ratio (OR) for exposure: if *a* are the participants with disease and exposure, *b* are those with no disease and exposure, *c* are those with disease and no exposure, and *d* are those with no disease and exposure, then the OR is $(a/c)/(b/d)$. In this example, the OR provides evidence for or against the hypothesis that consuming burned toast causes bowel cancer. As one can see, obtaining this evidence not only requires measuring the participants' health states and eating habits but also using one of the statistical techniques regularly employed by epidemiologists. The evidence provided by epidemiological studies, then, is theory-laden both in the ways Kuhn identified and with strategic theory in the form of study design and statistical techniques.

THEORY-LADENNESS AND THE EVIDENCE HIERARCHY

EBM highlights (at least) four kinds of theory-ladenness in evidence: the three outlined by Kuhn regarding observation and measurement, and strategic theory-ladenness, which pertains to study design and/or data interpretation postmeasurement. The best evidence requires the good forms of theory to be maximized and the bad forms to be minimized. As we can see below, the evidence hierarchy in EBM is structured according to these principles.

The Evidence Hierarchy

Important to EBM, and to the link between evidence and measurement presented in this chapter, is the hierarchy of evidence employed by evidence-based practitioners. As stated earlier, at the top of the hierarchy sit large-scale RCTs; weighted second are other kinds of epidemiological studies, such as cohort and case control observational studies; of third importance is pathophysiological/mechanistic reasoning; and finally, clinical experience and intuition (Howick 2011).

Clinical Experience and Intuition

"Clinical experience and intuition" is the weakest form of evidence to be found in the evidence hierarchy employed by practitioners of EBM. Nevertheless, even the evidence-based practitioner grants that decisions based on clinical experience amount to more than mere guesswork; that is, intuition and experience can provide at least some evidence to justify clinical decisions even if it is weak relative to those forms further up the hierarchy. Medical students observe many hundreds of patients, with almost as many pathological conditions. By observing their more senior colleagues' work, medical students directly experience whether a treatment was effective and if not, how the disease course developed and whether an alternative treatment was later successful. These experiences continue once the student qualifies and goes into practice herself. Eventually, by drawing on her clinical experience, in addition to her theoretical training at medical school, the doctor becomes skilled at recognizing physiological similarities between present and past patients and can thus identify pathological conditions and treatments more quickly and accurately. In other words, clinical experience, combined with the intuition necessary to liken present to past cases, better enables clinicians to form diagnostic, prognostic, and interventional hypotheses for their current patients.

But when a decision is made *purely* based on clinical experience and intuition, it is highly susceptible to implicit bias and false beliefs. This is partly because (i) relative to decisions based on clinical trials, decisions based on clinical experience and intuition rely on the observations of just a few patients (i.e., making the evidence less reliable) and (ii) this evidence is entirely devoid of good strategic theory-ladenness; that is, the statistical techniques used in epidemiological studies to ensure consistent treatments play no role in intuition. As a result, when making clinical decisions, neither the clinicians' prejudices (whether implicit or explicit) nor the differences between clinicians' intuitions or favored treatment techniques are constrained by clinical guidelines and best practices.

Pathophysiological Reasoning

Pathophysiological reasoning involves listening to a patient's complaints, examining the physiology of that patient, and then using one's understanding of the human pathophysiology to determine the appropriate treatment(s).

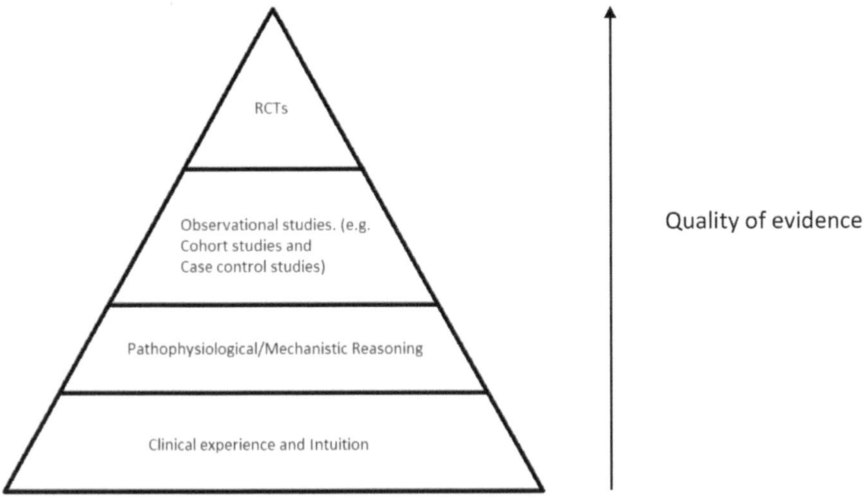

Figure 1.1 The Evidence Hierarchy.

Pathophysiological reasoning requires the kind of theory-ladenness typically associated with scientific observation. In treating a patient, first one must take her temperature, measure her blood pressure, examine her rash, and so on; second, one must apply these observations/measurements to the best understanding of human physiology to ascertain the appropriate intervention(s). Although there is much good theory-ladenness in the form of medical theory, history has shown that our understanding of human pathophysiology is often mistaken. Furthermore, like clinical experience and intuition, this form of evidence involves no good strategic theory-ladenness, so bad theory, such as implicit bias, can still lead to perceptual theory loading and issues with saliency.

Observational Studies

Public health professionals conduct several different kinds of observational studies. These studies broadly fit into the categories of the case-control study and the cohort study. For the sake of simplicity, in this section I focus on only a very basic case-control study.

Case-control studies generally involve comparing one cohort of participants from the population of interest with a suitable control group and/or alternative treatment group(s) to infer information about the etiology of a condition and/or the effectiveness or relative effectiveness of an intervention.

Suppose a case-control study shows the absolute risk of lung cancer (i.e., the percentage of the population who develop lung cancer over a specified period) given regular alcohol consumption (where this is a well-defined concept) to be significantly higher than the absolute risk of lung cancer given no alcohol consumption. In other words, more heavy drinkers get lung cancer during the ten-year study than nondrinkers. Were one to equate positive correlation with causation, then the study indicates a causal connection between drinking and lung cancer. But one cannot draw such a causal conclusion from that information alone since correlation does not imply causation.

Given our existing causal knowledge, alternative explanations for the correlation between drinking and lung cancer observed are not hard to find. For instance, the incidence rate of smokers in a population of heavy drinkers is far higher than that of a population of nondrinkers (i.e., more drinkers smoke than nondrinkers), and we already know of the strong causal connection between smoking and lung cancer. Now, it may still be the case that alcohol is a cause of lung cancer (i.e., that smoking is a confounder does not prove that alcohol has no causal effect), but given that smoking is clearly a strong confounding factor, simply determining differences between the absolute risk of lung cancer given drinking and the absolute risk of lung cancer given not drinking is insufficient to establish causation. It is clear, then, that any study attempting to calculate the causal effect of alcohol on lung cancer must, in one way or another, adjust for the confounding effect of smoking (e.g., Smart 2016, 71; Zang and Wynder 2001).

Nevertheless, for the evidence-based practitioner, observational studies are a much better form of evidence than intuition, clinical experience, and pathophysiological reasoning. Large-scale studies of this type can inform clinical guidelines and best practices in such a way that practitioners can standardize their clinical decisions and minimize the effect of implicit biases and differences between clinicians' intuitions. Observational studies are also rich with good strategic theory-ladenness. The study designs and statistical methods employed by epidemiologists set out to minimize the effects of confounding wherever possible, providing practical information concerning the causes and cures of disease. But the lack of randomization means that confounding is more likely with observational studies than with RCTs, so it is not the gold standard of evidence in EBM.

Randomized Controlled Trials

Crudely speaking, clinical trials are experiments designed to test the health implications of interventions. They attempt to identify the causes of disease and to test both the efficacy and effectiveness of curative and preventive interventions by applying well-specified interventions to a target population and assessing the outcomes. In the case of RCTs, this is typically achieved by comparing the outcomes of two groups of individuals. Each group is randomly drawn from the target population, and one is exposed to the intervention. After a time specified prior to the trial, the relevant qualities of the intervention and control groups are measured and compared so that, in theory, the causal effect of the intervention on the target population can be precisely calculated.

The literature on RCTs and their limitations is extensive, so a detailed discussion is well beyond the scope of this chapter.[7] But it is certainly worth highlighting some of the benefits of RCTs over observational studies. As we saw above, one of the most troublesome, if not *the* most troublesome aspect of any epidemiological study aiming to identify either the determinants of a disease or effective preventive or curative interventions is the presence of confounders (whether known or unknown). RCTs are designed to eliminate any such problems through the process of randomization. There are several ways in which randomization is achieved, and which strategy is used is largely dependent on the specifics of the background conditions, the interventions under scrutiny, and the population of interest.

Simple Randomization

The simplest method is to assign each member of the target population a number and then to use a random number generator to determine which individuals are to be in the intervention group and which in the control group. This strategy is designed to ensure that all possible confounding factors are equally distributed among the intervention and control groups and thus any differences found between the groups at the end of the study must be a product of the intervention under investigation. The RCT, then, deals with (or at least, it is designed to deal with) the problem of confounding far more easily than is possible with observational studies. This simplest method of randomization is rarely used in isolation, however. More often, one of the following strategies is employed.

Block Randomization

Block randomization determines the intervention and control groups by randomizing individuals within blocks in such a way that the same number of participants is allocated to each treatment under investigation; that is, within each block, if there is one intervention group and one control group, then half the participants in each group are "intervention participants" and half are "control participants." Within a block of four participants, there are six different possible orderings of patients. Where *a*, *b*, *c*, and *d* are patients; I is "group 1" and C is "group 2"; and I(*n*) allocates patient *n* to group 1, and C(*n*) allocates patient *n* to group 2, then the orderings are as follows:

- I(*a*) I(*b*) C(*c*) C(*d*);
- C(*a*) C(*b*) I(*c*) I(*d*);
- I(*a*) C(*b*) I(*c*) C(*d*);
- I(*a*) C(*b*) C(*c*) I(*d*);
- C(*a*) I(*b*) C(*c*) I(*d*);
- C(*a*) I(*b*) I(*c*) C(*d*)

"Allocation proceeds by randomly selecting one of the orderings," which is generally done using a random number generator (Efird 2011).

Stratified Randomization

Simple randomization, by definition, leaves the allocation of participants to treatment and control groups purely to chance. But simple randomization is often deemed inadequate since it leaves studies vulnerable to accidental bias. If there are known confounders, it usually makes sense to ensure that individuals subject to them are evenly distributed. *Stratified randomization* involves identifying specific covariates with potential influence on the dependent variable, to achieve balance between the case and control groups regarding the participants' baseline characteristics. Randomization in this way involves creating blocks for each combination of covariates and assigning each participant to the block relevant to them. Then, simple randomization (usually via a random number generator) is used to select members from each block to assign to the case and control groups (see Kernan et al. 1999).

Minimized Randomization

Minimized randomization is used when simple randomization is unlikely to produce balanced groups generally because of small sample sizes. Suppose one wished to measure the causal effect of cigar smoking on lung cancer. First, several stratification categories are selected as relevant to the study (e.g., sex, age over/under fifty, and cigarette smoker/nonsmoker). As well as being a cigar smoker, each participant satisfies some value for each. Take the first participant in group A to be a sixty-year-old male noncigarette smoker and the first participant in group B, to be a forty-year-old female non-cigarette smoker. If the next participant selected from the target population is a second cigarette-smoking male over the age of sixty, then one allocates him to group B, to "even things out." A fourth criterion of "number of participants allocated to each group thus far" must also be considered. Thus, at the end of the process, one should have two groups with roughly the same number of participants over sixty, roughly the same number of participants who smoke cigarettes and cigars, and roughly the same number of women (e.g., Scott et al. 2002).

These four distinct methods of randomization are all designed to eliminate, as much as possible, the effect of confounding on epidemiological studies. But although the criteria for choosing the randomization strategy to be employed are not arbitrary (some strategies are suitable for large populations, some are targeted at small samples, some cannot be used when confounders are known, and so on), the method used is ultimately chosen by those designing the study. One must question, then, how one can be certain that the randomization strategy employed is the most appropriate. Further, even if one can know that the most appropriate strategy on offer has been chosen, one cannot know whether better randomization techniques are yet to be discovered and accepted by epidemiology. Even if the best of all possible randomization techniques *is* used, there is no guarantee that the confounding effects on an RCT have been eliminated. After all, stratified randomization, for example, requires one to evenly distribute confounders, yet (i) it may be impossible to do this and (ii) some confounders may be overlooked/unknown. Both (i) and (ii) would be likely to skew the results of an RCT.

That one of several randomization strategies must be selected for RCTs to operate and that for each study, some strategies are better suited than others goes to show that RCTs are heavily theory-laden over and above the theory one identifies with observation and measurement.[8] Nevertheless, randomization undoubtedly improves one's chances of establishing genuine causal connections. It is this good strategic theory-ladenness that minimizes the possibility of implicit bias and inconsistency in clinical decision making, and places RCTs at the top of the evidence hierarchy.

EBM AND EVIDENCE AND MEASUREMENT

In the sections Evidence-Based Medicine, Patient-Centered Care, and Implicit Bias, I discussed the theory-ladenness of observation and the various guises in which it comes. As we saw, the history of science is full of examples supporting Kuhn's claims in this regard, and in cases such as measuring velocity or mass, the theory-ladenness of observation is plain to see. It is unsurprising, then, that in applying to medicine the concepts of perceptual theory loading, semantical theory loading, and saliency in observation, one can demonstrate how both good and bad theory-ladenness feed into medical theory and clinical decision making.

Even though the studies conducted by Green et al. (2007) and Schulman et al. (1999) outlined in this chapter indicate that *explicit* biases generally do not affect clinical decisions, bad theory, such as *implicit* bias, does play a role in the decision making process, leading to poorer health care for socially stigmatized groups. EBM sets out to eliminate this problem. To achieve this goal, as well as the forms of theory-ladenness that inevitably come from using measurement devices, such as thermometers and MRI scanners, the evidence prioritized by EBM employs the good strategic theory-ladenness that underlies clinical trials.

The structure of EBM's evidence hierarchy, which plays an important role in determining what "the best evidence" is, thus emphasizes the impact of different forms of theory-ladenness in evidence. The paradigm shift toward measurement that characterizes EBM brings into focus the theory-ladenness of both measurement and evidence, calling attention to how the theory-ladenness Kuhn identifies as inherent to all scientific measurement is similarly inherent in evidence. Furthermore, the EBM evidence hierarchy indicates how the many kinds of theory-ladenness, both good and bad, feature in determining the quality of evidence in science.[9]

NOTES

1. Note that I am not suggesting that practitioners of all forms of non-Western medicine are "swindlers." By practitioners of Western medicine, I mean qualified general practitioners (GPs), surgeons, pathologists of the kind found in hospitals in the West, and so on. I exclude all "alternative" or "traditional" forms of medicine, such as African sangomas and homeopaths. To emphasize here, I make no value judgments concerning the general effectiveness of non-Western medicine.

2. Note that outside of a clinical environment, acting in accordance with stereotypes is often necessary to behave rationally. Miranda Fricker (2007, 32) writes that when we imagine listeners being, "confronted with the immediate task of gauging how likely it is that what a speaker has said is true. Barring a wealth of personal knowledge of the speaker as an individual, such a judgement of credibility must reflect some kind of social generalization about the epistemic trustworthiness, i.e. the competence and sincerity, of people of the speaker's social type." Although Green et al.'s (2007) study showed no significant results linking explicit bias and clinical decision making, inevitably the explicit biases of clinicians will affect behavior toward some patients in some ways. Thus, even if for the most part clinicians can put their explicit biases to one side within a clinical setting, this is unlikely to always be the case.

3. At least not at this stage in the process. It might be shown that implicit biases play a role in the formation of the guidelines; in this chapter, however, I am more concerned with what EBM hopes to achieve than what it in fact does achieve.

4. Note that the fact that EBM eliminates implicit bias does not prove it to be a better form of medical practice than individualized, or patient-centered, care. However, it certainly seems to be a benefit of evidence-based practice.

5. When there is an objective value for some variable (e.g., the actual mass of an object), a measurement outcome becomes more accurate as it converges on this value. Whether and how this can be tested is beyond the scope of this chapter.

6. In the cases above, the skin color of the participants.

7. See Broadbent (2013) for an excellent philosophical examination of clinical trials.

8. In the strategic theory-ladenness sense.

9. Acknowledgments: I would like to thank Leah McClimans, Alexander Bird, Havi Cavel, Karim Thebault, Pendaren Roberts, Alex Broadbent, Veli Mittova, Chadwin Harris, and those who attended the Philosophy of Medicine Roundtable symposium at the 2016 PSA conference in Atlanta.

REFERENCES

Bensing, Jozien. 2000. "Bridging the Gap: The Separate Worlds of Evidence-Based Medicine and Patient-Centered Medicine." *Patient Education and Counselling* 39: 17–25.

Bogen, Jim. 2013. "Theory and Observation in Science." In *The Stanford Encyclopedia of Philosophy* (Summer 2014 Edition), edited by Edward N. Zalta. https://plato.stanford.edu/archives/sum2014/entries/science-theory-observation/.

Broadbent, Alex B. 2013. *Philosophy of Epidemiology*. London: Palgrave Macmillan.

Brownstein, Michael. 2015. "Implicit Bias." In *The Stanford Encyclopedia of Philosophy* (Winter 2016 Edition), edited by Edward N. Zalta. https://plato.stanford.edu/archives/win2016/entries/implicit-bias/.

Cartwright, Nancy. 2007. "Are RCTs the Gold Standard?" *BioSocieties* 2: 11–20.

Cooper, Rachel. 2004. "What Is Wrong with the DSM?" *History of Psychiatry* 15: 5–25.

Efird, Jimmy. 2011. "Blocked Randomization with Randomly Selected Block Sizes." *International Journal of Environmental Research and Public Health* 8: 15–20.

Fernández-Dols, José Miguel, and James M. Carroll. 1997. "Is the Meaning Perceived in Facial Expression Independent of Its Context?" In *The Psychology of Facial Expression*, edited by James A. Russell and José Miguel Fernández-Dols, 275–94. Cambridge: Cambridge University Press.

Fricker, Miranda. 2007. *Epistemic Injustice: Power and the Ethics of Knowing*. Oxford: Oxford University Press.

Green, Alexander R., Dana R. Carney, Daniel J. Pallin, Long H. Ngo, Kristal L. Raymond, Lisa I. Iezzoni, and Mahzarin R. Banaji. 2007. "Implicit Bias among Physicians and Its Thrombolysis Decisions for Black and White Patients." *Society of General Internal Medicine* 22: 1231–38.

Guyatt, Gordon H. 1991. "Evidence-Based Medicine." *American College of Physicians Journal Club* 114: A16.

Howick, Jeremy H. 2011. *The Philosophy of Evidence-Based Medicine*. Hoboken, NJ: John Wiley & Sons.

Karthikeyan, Ganesan, and Prem Pais. 2010. "Clinical Judgement and Evidence-Based Medicine: Time for Reconciliation." *The Indian Journal of Medical Research* 132: 623.

Kernan, Walter N., Catherine M. Viscoli, Robert W. Makuch, Lawrence M. Brass, and Ralph I. Horwitz. 1999. "Stratified Randomization for Clinical Trials." *Journal of Clinical Epidemiology* 52: 19–26.

Kuhn, Thomas S. 1996. *The Structure of Scientific Revolutions* (3rd edition). Chicago: University of Chicago Press.

McClimans, Leah. 2013. "The Role of Measurement in Establishing Evidence." *Journal of Medicine and Philosophy* 38: 520–38. doi:10.1093/jmp/jht041.

Rothman, Kenneth J., Sander Greenland, and Timothy L. Lash. 2008. *Modern Epidemiology* (3rd Edition). Philadelphia: Lippincott Williams & Wilkins.

Sackett, David L., William M. C. Rosenberg, J. A. Muir Gray, R. Brian Haynes, and W. Scott Richardson. 1996. "Evidence Based Medicine: What It Is and What It Isn't." *The British Medical Journal* 312: 71.

Schulman, Kevin A., Jesse A. Berlin, William Harless, Jon F. Kerner, Shyrl Sistrunk, Bernard J. Gersh, Ross Dube, Christopher K. Taleghani, John Burke, Sankey Williams, John M. Eisenberg, José J. Escarce, and William Ayers. 1999. "The Effect of Race and Sex on Physicians' Recommendations for Cardiac Catheterization." *New England Journal of Medicine* 340: 618–26.

Scott, Neil W., Gladys C. McPherson, Craig R. Ramsay, and Marion K. Campbell. 2002. "The Method of Minimization for Allocation to Clinical Trials: A Review." *Control Clinical Trials* 23: 662–74.

Smart, Benjamin T. H. 2016. *Concepts and Causes in the Philosophy of Disease.* London: Palgrave Macmillan.

Stepanikova, Irena. 2012. "Racial-Ethnic Biases, Time Pressure, and Medical Decisions." *Journal of Health and Social Behaviour* 52: 329–43.

Williams, Hywel, and Paul Seed. 1993. "Inadequate Size of 'Negative' Clinical Trials in Dermatology." *British Journal of Dermatology* 128: 317–26.

Zang, Edith, and Ernst Wynder. 2001. "Reevaluation of the Confounding Effect of Cigarette Smoking on the Relationship between Alcohol Use and Lung Cancer Risk, with Larynx Cancer Used as a Positive Control." *Preventive Medicine* 32: 359–70.

Chapter 2

Can Causation Be Quantitatively Measured?

Alex Broadbent

Since the middle of the twentieth century, the medical sciences have undergone a conceptual and practical revolution. The idea of measuring the effectiveness of medical interventions has taken hold. The idea comes from epidemiology, which initially focused on identifying the causes of disease and—not equivalently, but relatedly—detecting the effects of possibly harmful exposures. The idea was extended to measuring the effectiveness of potentially beneficial medical interventions and has come to dominate contemporary medical thought. The evidence-based medicine (EBM) movement goes so far as to suggest that the best or perhaps only credible evidence for effectiveness must be of a kind that quantitatively measures the effect of an intervention in a population. But even if one does not accept this priority claim, the importance of quantitative measures of "causal effect" as a kind of evidence in contemporary medicine is undeniable.

At the bottom of all this is a remarkable assumption, which is that it makes sense to quantitatively measure the effect that some medical intervention has in a population. You may think that "measure" sounds unduly strong; studies produce estimates, and "measure" sounds more certain than that. But if we assume that there can be facts about the unobservable, then there can be a measure of something where "measure" is a noun having the sense of a metric or a scale, even if one cannot actually or even possibly measure it, whatever exactly the latter verb means. And moreover, an estimate implies that there exists a measure, a metric, or a scale, even if one is unable to directly "measure" the item in question against that measure/scale/metric. If there were no measure, metric, or scale of length, just as there is probably no measure of general cosmic significance, then it would make no sense to estimate the length of a table, just as there is probably no sense in trying to give a quantitative estimate of the general cosmic significance of the table.

My question in this chapter is whether it makes sense to think of causation as having a quantitative measure, at least in the domain that matters to medicine: that of quantitatively expressing the effect of either hopefully therapeutic interventions or potentially harmful exposures on populations. I am not concerned here with the well-known methodological problems afflicting various kinds of study, whether observational or experimental, nor with the well-known difficulty of taking a result that has been found to hold for a study population and applying it to a target population. I am generally not concerned with epistemological matters in this paper. I am not concerned, for example, with questions such as how large a relative risk must be before we can consider making a causal inference (a poor question, but I am using it to set up a contrast). I am concerned with whether a relative risk can have a causal interpretation and if so, what it means.

This is interesting because philosophical theories of causation typically take a nonquantitative form and typically focus on specific, individual, or what medical scientists and statisticians sometimes call "actual" causation. Philosophers paradigmatically offer analyses of the form "*c* causes *e* if and only if," where *c* and *e* are individual, actual events. Where they do offer analyses of "general" causation, these analyses are likewise usually nonquantitative, offering conditions for events of type *C* to cause events of type *E*.

Contrast common epidemiological expressions, such as relative risk = 20 or population attributable fraction = 65%. These measures on their own do not imply that a causal inference has been made, but where one has been made, they can be used to say something about the result of that inference: to quantify it. This is particularly apparent in measures of attributability, which really make no sense if they are not at least potentially interpretable as making causal claims. But what do these causal claims assert?

Philosophers have not addressed this question, as I shall argue below. And in recent years, the question has driven a debate within contemporary epidemiology, in which I have been involved, concerning a movement known as the *Potential Outcomes Approach* (POA). The POA seeks to impose strict conditions on the meaningfulness of measures of causal effect.

In the section that follows, I will start by saying more about the idea of estimating causal effect and the POA critique, before considering in the section "Philosophy and General Causation" what philosophical theories might offer to help. Sadly, the answer is, not much. In "Preemption and the Potential Outcomes Approach," I then consider the problems that the POA view encounters in relation to preemption before proposing an alternative theory in the section on "Quantitative Causal Claims as Counting Mechanisms."

ESTIMATING "CAUSAL EFFECT"

One textbook goal for epidemiological research is "quantification of the causal relation between exposure and outcome" (Savitz 2003, 9). A source of conceptual difficulty is that the measures used for this quantification are, in the first instance, measures of association. They become measures of causation if the association in question is causal. This is rough; making it more precise is the point of this chapter.

For example, relative risk (or risk ratio) is the risk among an exposed group divided by risk among an unexposed group:

$$RR = \frac{R_E}{R_U}$$

(Risk is the number of outcomes divided by population size at the start of the period.) Population attributable risk is the difference between total population risk and risk in an unexposed group divided by the total population risk:

$$PAR = \frac{R_T - R_U}{R_T}$$

Clearly, one cannot "attribute" a risk in any normal sense merely by performing the above simple arithmetic. The question of whether it is correct to interpret a measure such as RR or PAR causally thus has two parts: an epistemological and a semantic one. The epistemological part receives more of the attention; fundamentally, it concerns whether there are other possible explanations for the observed associations besides the causal hypothesis. The semantic question concerns what meaning these measures have when causally interpreted, and this is the question for this chapter.

Recent developments in epidemiology have made the question pressing. The POA seeks to restrict the causal claims that epidemiologists should consider and the causal questions they should ask on the basis that the restriction sorts those that are meaningful from those that are not. The restriction was originally characterized in terms of *interventions*, with one proposal being that causal questions are well defined when interventions are well specified. Difficulty with a clear and plausible definition of "intervention" and the unfortunate connotation of that term linking it to humanly feasible activities have led some to pull back from this expression. But the core of the idea remains that to meaningfully express a measure of causal effect, one must specify some counterfactual contrast, reasonably precisely, against which the effect in question is measured (VanderWeele 2017).

The restriction is wonderfully motivated in a paper by Miguel Hernán and Sarah Taubman (2008), who tell the following story. A benevolent but entirely autocratic king was looking for ways to improve the health of his subjects, and he wondered about the effect of obesity on mortality. He called some scientists together and asked them to investigate using any resources they needed. The scientists performed a randomized controlled trial (RCT), which they saw as the gold standard in medical research. They randomly assigned 1 million people each to three intervention groups, one testing the effect of a restricted calorie, low carbohydrate diet; one testing the effect of an hour of strenuous exercise per day; and one testing the effect of both. They measured the reduction in body mass index (BMI) over a thirty-year period and compared mortality over that period to that of the general population.

The results were as follows. The diet group saw a reduction of 50,000 deaths annually compared to the general population. The exercise group saw a reduction of 100,000, and the group where both interventions were run saw a reduction of 120,000 deaths. Yet as it happens, reduction of BMI was the same between the three groups; in other words, they each lost the same amount of weight, but the effect on their mortality differed.

If you think this is implausible, bear in mind that this is a thought experiment. The point Hernán and Taubman are making is that in this situation, the king might be rather irritated. He might ask, "So, after thirty years of unlimited resources, tell me: Have you found out how much mortality obesity causes?" And the investigators would have to reply, "Well, no, we can't answer that question. It seems to depend on how obesity is reduced. We need to do further work to understand it."

Now, imagine a neighboring country ruled by a president who is bound by a constitution and who is envious of the remarkable project in the neighboring kingdom. Some similar initiative could win him many votes in an upcoming election, and besides being prevented from ordering compulsory trial participation by constitutional issues (which perhaps could be finessed), he does not have thirty years to squander waiting for the results. So he calls in a data analyst and asks whether she can help him if she has access to the country's excellent, prospectively recorded electronic health records. She spends two weeks on the problem and returns with the verdict that 150,000 excess deaths each year are attributable to obesity.

At first glance, it looks like the data analyst has achieved something that the trialists could not: she has answered the question "How much mortality does obesity cause?" This is a quantity that to the king's frustration, the trialists cannot estimate. However, argue Hernán and Taubman, the apparent advantage reverses when the question becomes "What should we do?" The trialists can say "That's easy; promote a combination of diet and exercise since that yields the largest drop in mortality." The data analyst, however, can only say "Reduce obesity." There are many ways to do this and not all of them improve mortality; some, like smoking, are positively harmful.

Hernán and Taubman are making several points with this story. One of these points concerns the usefulness of estimates of measures of causal effect. I will not deal with that here. Another concerns the *meaningfulness* of such measures. The data analyst's inability to recommend a specific course of action is not only supposed to illustrate that her estimate is not useful; it is also supposed to cast doubt on whether it is *meaningful* in the first place.

The idea appears to be that if the causal claim were meaningful, it would provide us with information about what would happen in some counterfactual scenario in which the causal exposure was different. The reasoning behind this idea, in turn, is that causal claims are characterized by carrying information about counterfactual scenarios. In short, the idea is that causal claims entail counterfactuals, that counterfactual dependence is necessary for causal claims, at least where those causal claims are quantitative estimates of causal effect. The story effectively poses a rhetorical question along the lines "What does it mean to attribute some quantity of outcome to exposure (mortality to obesity) if not that a reduced exposure would have led to that amount of reduction in outcome (mortality)?"

The paper confuses the counterfactual with the actual and asks what actual intervention would reduce mortality going forward. However, the point really concerns the meaningfulness of the attributability claim given that it appears not to yield a forward-looking prediction contingent on a contemplated intervention is a close cousin, of which the historic counterfactual is a close cousin, both semantically and epistemologically. This is not a point to quibble on.

What does deserve quibbling is the idea that counterfactual dependence is necessary for causal claims. This idea is almost universally rejected by philosophers (the only exception I know of being David Coady [2004]). But we should not be too quick to wield the bellows over the embers of the stock objections. As the epidemiological debate has progressed, the scope of this claim has been restricted. It has been acknowledged, in at least some writings, that counterfactual dependence is not necessary for causation in general (VanderWeele 2016). A distinction is drawn between *causal identification* and *estimation of causal effect.* This is a distinction between two kinds of *investigation* (the focus of the epidemiological debate being on epistemological matters), but it corresponds to a distinction between two kinds of *claims*, one asserting that *C* causes *E* and one making some sort of quantitative estimate concerning the causal relation between *C* and *E*.

Elsewhere (Broadbent, Vandenbroucke, and Pearce 2016: 1841) I have identified the tenets of the POA as the following:

1. Counterfactual dependence of *E* on *C* is not necessary for *C* to cause *E*, but it is sufficient (POA's basic metaphysical stance).
2. Sufficient evidential conditions currently exist for attributing the counterfactual dependence of *E* on *C*, but necessary conditions currently do not; the POA identifies some (but not all) of these sufficient conditions (POA's basic epistemological stance).
3. Causal inference includes two distinct aspects: *causal identification*, in which the truth value of a claim of the form "*C* causes *E*" is determined, and *quantitative causal estimation*, in which a numerical value *n* is estimated for a claim of the form "*C* has *n* effect on *E*" (the identification/estimation distinction).
4. Adequately well-defined counterfactual contrasts are necessary for giving meaning to quantitative estimates of causal effect (POA's semantic stance on estimation).

The claim that concerns us here is item 3. This is not simply the claim that counterfactual dependence is necessary for causation; it is the more refined claim that for an *estimate of causal effect* to be meaningful, there must be a clearly specified counterfactual scenario in which the cause is different in a sufficiently clearly defined way and against which the actual effect is measured. If the counterfactual scenario is not clearly enough specified, then the difference between that scenario and the actual world will also not have a definite value and thus cannot be "measured." Coming up with several deaths attributable to obesity, for example, is not a meaningful exercise if one has not sufficiently clearly specified the scenario in which obesity is absent. It is a bit like trying to measure the distance between a wall and a puddle without specifying whether the measurement is to be taken to the edge of the puddle, or its center, or some other part of it.

In Hernán and Taubman's paper, emphasis is laid on the fact that the data analyst cannot assert a counterfactual conditional about what would happen if obesity were less. This is not quite the same as saying that the counterfactual scenario specified by the data analyst is imprecisely specified. What explains this?

As far as I can see, the role of the counterfactual conditional in Hernán and Taubman's paper, and in POA thinking in general, is as a sort of test for precise specification of a counterfactual scenario. If one says that 150,000 deaths are attributable to obesity and one has not adequately specified the counterfactual scenario in which obesity is absent and these deaths do not occur, then one will not be able to say that if obesity had been absent, mortality would have been 150,000 less. This is because if the antecedent scenario is not sufficiently clearly specified, then we cannot say much about what would have happened in that scenario. In my reconstruction of the POA reasoning (no explicit argument of this kind is given), estimates of causal effects must support counterfactuals because the failure to do so shows inadequate specification of a counterfactual contrast and one cannot measure the difference between an inadequately specified scenario (a puddle) and anything else.

This, at least, is the argument as I infer it to be. It has two philosophically interesting consequences: that causal effect can be quantitatively measured and that it can be quantitatively measured only when a counterfactual contrast is adequately specified. The next two sections consider these consequences each in turn.

PHILOSOPHY AND GENERAL CAUSATION

In this section I briefly review what philosophers have offered by way of theories of causation that might help us assess the epidemiological assertion that some causal claims can be quantitatively expressed.

Philosophers have had disappointingly little to say about "general causation" and have even been rather unthoughtful about defining the object of analysis. The phrases "general causation," "causal generalizations," and "general causal claims" each means something different. The first phrase suggests a thing that exists at a "general" level, the second suggests a generalization about causal relations holding at a nongeneral level, and the third is ambiguous between the two.

Corresponding to the first two phrases, there are fundamentally two approaches to general causal claims: the reductionist and the antireductionist. An example of antireductionism is Ellery Eell's probabilistic theory of causation, which asserts that type-level and token-level causal claims are wholly distinct. Eell's is clear that a type-level claim, such as "Smoking causes lung cancer," can be true even if there are no tokens, which in this example could come about either because no actual smokers get lung cancer (even if everyone smokes) or because there are no actual smokers. As Eells (1991, 7) puts it,

> the way I shall understand statements of positive type level probabilistic causation, they do not even imply that any token of the cause type ever exemplified; nor do they even imply that if the cause type were universally exemplified then the effect type would have to be exemplified at least once. I will call this the *no-token thesis*, to indicate that there need be neither any token of the cause, nor any token of the causal relation, for the type-level causal relation to hold.

Eells's position is perhaps hard for him to avoid given the probabilistic nature of his account. The difficulty with it is that it is very implausible for a large class of causal claims that epidemiologists want to make. We can roughly distinguish two kinds of causal claims that epidemiology might be involved in investigating: those of causal *capacity* and those of causal *effect*. Eells's view ignores the second of these claims. Claims such as "Smallpox causes death" or "Sirens cause shipwrecks" are true if taken to express the capacity of smallpox or sirens to cause death or shipwrecks (Broadbent 2013, 35). But they are false if taken to express the effect that smallpox has on actual populations (since it has been eliminated) or the effect that sirens have on actual ships (modern vessels being so large that it takes them twenty miles to make a turn, by which time the sirens are out of earshot and the sailors are themselves again).

Epidemiologists do not merely investigate the capacity of exposures to cause outcomes; they also, as we have seen, seek to establish what effect these exposures have on actual populations. It is in this context that quantification is important because the assertion that an exposure has *some* effect rather than none at all is useless for practical purposes: size is everything in this context. Even the position of Neptune has some effect on the force I need to exert to raise my arm due to the gravitational force it exerts on my arm, but the size of the effect means I can ignore it when moving my arm.

In the context of claims of causal effect (as opposed to capacity), "Smoking causes lung cancer" means something like "Smoking *actually* causes *actual* lung cancer." For this to be true, there must be some actual lung cancer, and it must be caused by smoking, meaning that there must also be some smoking. For this kind of causal claim, which is the kind that epidemiologists hope to quantify, Eells's expression of antireductionism does not hold. In this sense, smallpox does *not* cause death, and sirens do *not* cause shipwrecks. And the intuitive acceptability of these statements shows that the epidemiologists are not working with some specialist notion; our normal general causal claims are also ambiguous between capacity and effect claims and are often intended as the latter.

Could one be an antireductionist about general causal claims and yet reject Eells's no-token thesis? Perhaps, although one would need to explain how the two kinds of causal claims were related. However, it seems to me that a reductionist view is in any case more attractive if the effort is to quantify causal effect. This is because the natural way to understand that quantification is the result of counting the instances of the causal relation that we are trying to quantify. For this reason, I am inclined toward a reductionist account of general causal claims, at least for claims of causal effect (as opposed to capacity).

Philosophers have not offered detailed reductionist theories of general causal claims perhaps because the term "causal generalization" suggests that there is no need for such a theory. If general causal claims are generalizations, then there is nothing more to be said about them. "Smoking causes lung cancer" means "All smoking causes lung cancer," and "causes" means whatever it means at the individual level, for example, that if person A had not smoked, she would not have gotten lung cancer (on a counterfactual theory of individual causation, such as Lewis's [1973]).

This approach is equally ill suited to epidemiological purposes. It is not the case that all smoking causes lung cancer—a point to which Eells is especially sensitive and which motivates his probabilistic theory and his no-token thesis in particular. Most smokers do not get lung cancer, which is a rare disease even among smokers. Moreover, universal quantification is not the sort of quantification that epidemiologists are interested in; the way philosophers use "quantification" is quite odd, in that they certainly do not mean to indicate anything that could be expressed with a number. Epidemiologists are interested in quantification, not in the all-or-something sense in which philosophers use the term, but in the numerical sense: they are interested in *quantitative* quantification.

There are no philosophical theories of quantitative causal claims as far as I know. Epidemiologists have beaten philosophers to it. If we read between the lines, it is possible to infer a theory of quantitative causal claims from the POA, for at least some kinds of causal claims, as follows:

> A certain quantity n of outcome/effect E is attributable to exposure/cause C (or to a certain quantity m of C) if and only if, were it the case that C was absent (or m less), E would be n less.

Let us call this the POA theory of quantitative causal claims. It is a simple, counterfactual theory, asserting that absence of exposure (or of a certain quantity of exposure—often epidemiologists want to measure the effect of elevated levels of an exposure, which is normally present in much lower levels) would lead to the absence of the effect in the amount indicated by the measure.

This theory concerns only attribution and not other kinds of measure, but measures of attributability seem a good place to start, being directly concerned with measuring causal effect. Let us now assess this theory.

PREEMPTION AND THE POTENTIAL OUTCOMES APPROACH

As discussed above, the POA asserts that causal effects can be estimated *only* when a corresponding counterfactual is true because (as I understand it) the falsity or meaninglessness of a counterfactual indicates imprecise specification of a counterfactual scenario, which therefore cannot form a reference point for any measure. This suggests an obvious potential source of difficulty for the POA and its theory of quantitative causal claims. Philosophers almost all agree that counterfactual dependence is not a necessary condition on causation at the individual level. The question then arises whether it can be a necessary condition on quantitative causal claims, especially (but not only) if one agrees with my suggestion that they are most naturally interpreted as reducible, in the sense of being numerical claims about individual instances of causation.

The reason that most philosophers reject counterfactual dependence as a necessary condition on individual causation is the very common scenario known as *preemption*. This occurs whenever something happens that would have happened anyway even if the actual cause had not occurred or had occurred but did not cause the effect. For example, suppose that one night I cook the dinner but if I hadn't cooked it, my wife would have. Then it's false that had I not cooked, there would have been no dinner; yet it's clearly true that I caused dinner to come about (Lipton 2004). We have causation without counterfactual dependence. Philosophers often illustrate preemption with entertaining and outlandish scenarios: two assassins aiming at the same president, and so forth. But this must not be allowed to obscure the fact that preemption is extremely common and is a feature of the causal relation that we make use of. Preemption is why cars have spare tires, why hospitals have backup generators, and why fighter pilots carry parachutes.

Let us now see whether preemption causes a problem for the POA theory of quantitative causal claims. Consider one of the arms of the king's trial, which ought to yield a well-defined causal claim. The trialists assert that 100,000 excess deaths are attributable to the absence of the exercise intervention that they tested—or, more naturally, that 100,000 fewer deaths occurred in the exercise group because of the exercise. On the POA theory of quantitative causal claims, this means that had the exercise and control groups been swapped—or, where the control is the general population, had a different group of a million been randomly selected from the general population—the findings would have been replicated.

This is exactly what trials seek to ascertain. There is even a trial design, the crossover trial, that is intended to establish exactly this: that the effect would have been the same in the control group had it been given the intervention. (Of course, crossover trials cannot easily be done where the outcome of interest is death, as in this case.) The POA appears to be a very plausible interpretation of the kind of quantitative causal claim that trials, at least, seek to assess.

There are, however, two problems that arise if the POA theory of quantitative causal claims is extended to claims that do not concern the result of interventions that are tested in trials, but the observed effect of actual exposures in populations. The first is that removing an exposure typically results in a smaller reduction in outcome than this theory would suggest. The reason for the lesser effect is that there is often something else that will bring the effect in question about and which is either correlated with the exposure (a confounder) or will become an exposure when the original exposure is removed (for example, if some cigarette smokers switch to cigars when they give up, or drink more soft drinks, or eat more, or get less sleep, etc.). In other words, preemption occurs, at the individual level, in some cases. More exactly, within the actual effect are several individual cases where the actual cause preempts some other event; in the counterfactual scenario where the actual cause is absent, that event would occur and would bring about the effect in question in the individual.

At first, this looks like a practical wrinkle that does not pose a theoretical problem. The POA advocate can insist that, strictly speaking, the counterfactual scenario needs to be specified tightly enough to exclude the operation of backup causes. This eliminates any preemptors from among the individual cases of causation in your actual sample. Then, the counterfactual in question will be true: removing the cause *while specifying that the backups do not operate* will result in a reduction of the effect by the amount indicated in the measure.

But this response is not wholly satisfactory. What we wanted was a theory of the meaning of the quantitative causal claims that epidemiologists make when they make causal attributions. This is not what we have been given. The POA theory of quantitative causal claims does not express the meaning of the quantitative causal claims made in many practical contexts but only those made in the rather contrived scenario of expressing the results of a trial, along with any others that can approximate those claims. Hence, the emphasis on well-specified interventions: the idea is to get observational studies to approximate clinical trials as closely as possible, with the meaningfulness of the results depending on the degree of approximation. This is a little backward; it would be nice to see whether we could work out the meaning of the results of observational studies that do not approximate clinical trials, if possible, before deciding that they have no meaning.

The second problem is that no matter how tightly a counterfactual scenario is specified at the population level, preemption may still occur at the individual level. The trouble is that an exposure itself may be a preemptor among those individuals who would have suffered the outcome anyway. For example, among smokers who get lung cancer is a tiny number who would have gotten lung cancer anyway. It may be that smoking is a cause of their lung cancer too—either that it somehow "displaces" the cause that otherwise would have operated or that it works alongside that cause. The point can also be brought out by considering the possibility of exposures that *prevent* the outcome in some people but bring it about in others. Perhaps exercise will make many people healthier, but for some—perhaps sedentary persons with undetected aneurisms—it may cause death. The net effect of exercise on the population will consider the tiny number of deaths caused by exercise, which in this scenario is in effect swapping around a few of the would-be survivors in the population, as well as benefiting a large number in a straightforward way. You might count 100,000 fewer deaths than in the control group and attribute this to the carefully specified exercise intervention, and you might be right that if that group had not exercised, there would have been 100,000 more deaths. But exercise might in fact have operated to prevent more deaths than this if it also caused some deaths.

An example I have used elsewhere to make the same point is a fictitious population of Himalayan porters who carry heavy loads, some of whom suffer from herniated discs (Broadbent 2011; Broadbent 2013). Suppose that a certain number of these herniated discs are attributed to carrying heavy loads; in the POA framework, this is equivalent to saying that the remainder would have occurred anyway. The trouble is that even among those that would have occurred anyway, it may be that carrying heavy loads is causal: it contributes to the actual herniation even if a counterfactual herniation would have occurred in the absence of heavy loads. If this is so (and this is all hypothetical, for illustration), then the counterfactual does not accurately represent the amount of causation that is present, to put it crudely. The heavy loads may be causal in *every* case of herniated disc among the porters even if some herniation would have occurred if they had not been porters but rather office workers.

This point is also drawn from the epidemiological literature, where it has been pressed by Sander Greenland particularly in the context of incorrect usages of epidemiological evidence in legal proceedings (Beyea and Greenland 1999; Greenland 1999; Greenland and Robins 1988; Robins and Greenland 1989). The same point seems to apply to the POA account of quantitative causal claims if those claims are taken as quantifying individual cases of causation. This may be denied, of course: the POA theory could be taken as an antireductionist account, quantifying causal effect as a sui generis phenomenon (or at any rate one not reducible to individual causation). I am not going to attempt to refute that suggestion here; it is an interesting possibility. However, I think it is not one that POA advocates would in fact be drawn to; the idea that causation is irreducible is not in line with the attempt to better understand mediation and interaction, which drives much of that literature. In any case, a proper elaboration and defense of quantitative causal claims would need to be offered if it were accepted that they are not claims about the number of individual cases of causation.

QUANTITATIVE CAUSAL CLAIMS AS COUNTING MECHANISMS

The most obvious way to understand quantitative causal claims, in my view, is as claims about the number of individual cases of causation. The existence of preemption means that the number of individual cases of causation is not the same as the net difference that a cause makes in a population, no matter how tightly specified the counterfactual scenario we compare it to. If we want a reductionist account of quantitative causal claims, we need a way of getting around the fact that the net difference the cause makes—which is what epidemiology measures—does not necessarily equal the total effect of the cause.

I suggest that the simplest solution to this difficulty has two parts. First, attributions must be understood as having the form of an inequality: to say that 100,000 excess deaths are caused by inactivity is to say that *at least* 100,000 excess deaths are caused by inactivity. The measure is a measure of a lower bound: the near edge of the puddle. Second, the counterfactual element of the theory must be dropped in favor of something that makes a direct existential claim about causal relations at the individual level (and here I diverge from what I have suggested previously [Broadbent 2013, 50–55]). I suggest we employ the concept of a mechanism. Attributions should be understood as asserting that in at least this many individual cases, a mechanism was instantiated by which the cause led to the effect occurring.

The reason for this is that preemption may occur even among those cases that are detected in the net effect (the first problem identified above). But in a nonmysterious epidemiology, what we are presumably saying with causal attributions is that something happened, by which the cause led to the effect in at least a certain number of cases—which we have counted; we cannot guarantee that if the cause is removed, the effect will not happen for some other reason, but we can assert that in at least this many cases, it happened because of the cause. The point of invoking the concept of mechanism is to emphasize the reality of the individual causal relation and identify it as something that presumably calls for further unpacking; also, because in many practical contexts, forgetting to ask whether a plausible mechanism exists to connect cause and effect can lead to wild and implausible effect estimates. To summarize, a certain quantity n of outcome/effect E is attributable to exposure/cause C (or to a certain quantity m of C) if and only if, in at least n cases of E, a mechanism operates by which C causes E. This gives us a reduction of quantitative causal claims to individual cases of causation.

CONCLUSION

I have sought to make sense of quantitative measures of causal effect in epidemiology. Extant philosophical work is not much help. The front-runner in the epidemiological literature, the POA, offers a clear account of the meaning of quantitative causal claims, but one that appears to direct us toward antireductionism, which I find undesirable and also consider to be at odds with the ambitions of the POA to better understand mediation and interaction (both of which imply understanding more about causal relations and not leaving them uninterrogated). I made a different proposal, which sees measures of causal effect as "at least" claims about the operation of mechanisms connecting instances of cause to effect.

REFERENCES

Beyea, Jan, and Sander Greenland. 1999. "The Importance of Specifying the Underlying Biologic Model in Estimating the Probability of Causation." *Health Physics* 76: 269–74.

Broadbent, Alex. 2011. "Epidemiological Evidence in Proof of Specific Causation." *Legal Theory* 17: 237–78. doi:10.1017/S1352325211000206.

Broadbent, Alex. 2013. *Philosophy of Epidemiology*. London and New York: Palgrave Macmillan.

Broadbent, Alex, Vandenbroucke, Jan P., and Pearce, Neil. 2016. "Formalism or Pluralism? A reply to commentaries on 'Causality and Causal Inference in Epidemiology.'" *International Journal of Epidemiology* 45(6): 1841–51.

Coady, David. 2004. "Preempting Preemption." In *Causation and Counterfactuals*, edited by John Collins, Ned Hall, and L. A. Paul, 325–40. Cambridge: MIT Press.

Eells, Ellery. 1991. *Probabilistic Causality*. Cambridge: Cambridge University Press.

Greenland, Sander. 1999. "Relation of Probability of Causation to Relative Risk and Doubling Dose: A Methodologic Error That Has Become a Social Problem." *American Journal of Public Health* 89: 1166–69.

Greenland, Sander, and James Robins. 1988. "Conceptual Problems in the Definition and Interpretation of Attributable Fractions." *American Journal of Epidemiology* 128: 1185–97.

Hernán, Miguel A., and Sarah L. Taubman. 2008. "Does Obesity Shorten Life? The Importance of Well-Defined Interventions to Answer Causal Questions." *International Journal of Obesity* 32: S8–S14.

Lewis, David. 1973. "Causation." *The Journal of Philosophy* 70: 556–67. http://www.jstor.org/stable/2025310.

Lipton, Peter. 2004. *Inference to the Best Explanation*. 2nd Edition. London and New York: Routledge.

Robins, James, and Sander Greenland. 1989. "The Probability of Causation under a Stochastic Model for Individual Risk." *Biometrics* 45: 1125–38.

Savitz, David A. 2003. *Interpreting Epidemiologic Evidence: Strategies for Study Design and Analysis*. Oxford: Oxford University Press.

VanderWeele, Tyler J. 2016. "On Causes, Causal Inference, and Potential Outcomes." *International Journal of Epidemiology*. doi: 10.1093/ije/dyw230.

Chapter 3

Absolute Measures of Effectiveness

Jacob Stegenga and Aaron Kenna

A central aim of medical research is causal inference. Does this drug have harmful side effects? Is this medical intervention effective? Does this chemical cause cancer? To provide evidence that bears on these important questions, many sorts of measurements are made in a variety of types of studies. These measurements generate a plethora of data, and these data must be quantitatively summarized so they are rendered relevant to causal hypotheses. That is, to render *measurements* made in medical research into *evidence* for a causal hypothesis, those measurements must be transformed into summary quantifications, called "outcome measures." This chapter has two aims. First, we argue for the superiority of one form of outcome measure, called absolute measures. Second, we argue against a widely held myth in epidemiology. The myth is that in observational methods, such as case-control studies, only the relative outcome measure called the odds ratio can be calculated, and we argue that there is no justification for this myth.

An outcome measure is an abstract, formal statement describing a relation between the value of a property in one group of a study and the value of that property in the other group. For example, a study might measure cholesterol levels in a group that received an experimental drug and in a group that received a placebo, and an outcome measure would compare the values of those cholesterol levels between the two groups. There are many possible outcome measures; we describe the most commonly used outcome measures in the subsection below titled "Outcome Measures." When values for measured properties are substituted into an outcome measure, the result is called an "effect size." This terminology is misleading because although these numerical summaries of measurements are relevant for causal inferences, and specifically for estimating the strength of a causal relation between an experimental intervention and a measured outcome, there is no straightforward relationship between an effect size and the strength of a causal relation.

There are outcome measures for both continuous and dichotomous properties. For both, the choice of outcome measure is important and can have significant influence on causal inferences. Our initial claim regarding the superiority of absolute outcome measures can be made simply by focusing on dichotomous parameters.

We begin below by defining the most commonly employed outcome measures in medical research. We define these measures in the standard way that medical scientists do, and we restate the definitions using conditional probabilities (this is useful for our later arguments). In the section "Absolute Measures Are Superior to Relative Measures," we offer two related arguments for the superiority of absolute measures over relative measures. The unwarranted myth that case-control studies cannot furnish absolute measures is our target in "The Myth of the Odds Ratio."

OUTCOME MEASURES

For dichotomous parameters, the most common outcome measures include the odds ratio, relative risk (sometimes called risk ratio), relative risk reduction, risk difference (sometimes called absolute risk reduction), and number needed to treat. To define these measures, we construct what is called a two-by-two table: we consider a hypothetical study that has an intervention group (E) composed of subjects who receive the intervention under investigation and a control group (C) composed of subjects who receive a control intervention (perhaps placebo, or a competitor intervention, or nothing at all), in which a binary outcome is either present (Y) or absent (N), where the number of subjects with each outcome in each group is represented by letters (a–d) (see Table 3.1).

Here are the most commonly used outcome measures defined regarding Table 3.1:

- Relative risk (RR) = [a/(a+b)] / [c/(c+d)]
- Relative risk reduction (RRR) = [[a/(a+b)] – [c/(c+d)]] / [c/(c+d)]
- Odds ratio (OR) = (a/c) / (b/d) = ad/bc
- Risk difference (RD) = a/(a+b) – c/(c+d)
- Number needed to treat (NNT) = 1 / [[a/(a+b)] – [c/(c+d)]]

Table 3.1 Two-by-two table for defining outcome measures

Group	*Outcome*	
	Y	N
E	a	b
C	c	d

Note a few aspects of these outcome measures. RR is simply the proportion of subjects in the experimental group with a Y outcome *divided* by the proportion of subjects in the control group with a Y outcome. RD, on the other hand, is the subtracted *difference* between these proportions. An equivalent expression for RRR is 1 – RR. Note too that NNT is the inverse of RD. RR and RRR are the most common relative outcome measures, and RD is the most important absolute outcome measure. OR is simply the proportion of Y-outcome subjects who received the intervention divided by the proportion of N-outcome subjects who received the intervention. It is standard to simplify the expression of OR as we have done.

Consider, for example, a hypothetical RCT evaluating the efficacy of a new intervention for the prevention of death against a placebo control (see Table 3.2).

In this hypothetical trial, we have the following effect sizes:

- RR = 0.12/0.20 = 0.60
- RRR = (0.20 – 0.12)/0.20 = 0.40
- RD = 0.20 – 0.12 = 0.08
- NNT = 1/0.08 = 12.5

Relative risk provides information regarding the efficacy of a treatment. We may conclude the following from the RR calculated in our example: If RR = 1, there is no difference in risk of mortality between the intervention and control groups. If RR > 1, the treatment increases the risk of mortality (for simplicity we assume statistical significance). And if RR < 1, the treatment decreases the risk of mortality.

RRR, on the other hand, provides information about the size of the relative change in risk associated with a treatment. In our example, RRR = .40, or 40 percent. Hence, we may report the clinical data as showing that patients who underwent the experimental treatment experienced a 40 percent reduction in the risk of mortality.

It is useful to define outcome measures in terms of conditional probabilities. The probability of a subject having a Y outcome given that the subject is in group E, P(Y|E), is a/(a+b), and likewise, the probability of having a Y outcome given that the subject is in group C, P(Y|C), is c/(c+d). Thus, for example, we have the following:

$$RR = P(Y|E)/P(Y|C)$$

$$RD = P(Y|E) - P(Y|C)$$

Table 3.2 Hypothetical data from a trial

	Survived	*Died*	*Probability of Death*
Intervention	88	12	12/(88+12) = .12
Placebo	80	20	20/(80+20) = .20

These formal statements are relatively uninformative without interpretation. Recall that the central use to which outcome measures are put is to aid in causal inference. Specifically, in this context there are three crucially important kinds of causal inferences that one might be interested in, corresponding to hypothesis types of differing degrees of generality. These hypothesis types are efficacy hypotheses, effectiveness hypotheses, and predictive hypotheses. Efficacy hypotheses are claims about a causal relation in the context of a medical study, such as "this drug lowered blood sugar levels in this trial." Effectiveness hypotheses are claims about this causal relation in a more general context (in the wild, so to speak), such as "this drug lowers blood sugar for middle-aged diabetic patients." Predictive hypotheses are future claims about this causal relation for a particular person, such as "this drug will lower my blood sugar." How outcome measures should be interpreted depends on these different hypothesis types.

To illustrate, we focus on RD. Since RD is a difference of conditional probabilities, it is itself a probability. What sort of probability is it? That depends on which hypothesis type is in question.

We use RD_{trial} to represent the risk difference that is measured in an RCT. Evidence from an RCT is directly relevant to efficacy hypotheses. RD_{trial} is an *actual frequency*: that which was in fact measured in the trial. If $RD_{trial} > 0$, then we have evidence for an efficacy hypothesis.

We use RD_{target} to represent the risk difference that would be expected if the intervention were to be used in the wild. RD_{target} is an *expected frequency*: that which we would expect to observe if the intervention were used in a general context (in the wild). Evidence from an RCT is less directly relevant to effectiveness hypotheses than it is to efficacy hypotheses, but it is nonetheless relevant. The aim is to make an inference such as $RD_{trial} \longrightarrow RD_{target}$. For a variety of reasons that we do not articulate here, we do not think that $RD_{trial} = RD_{target}$ or that this inference is deductive. In typical cases, RD_{target} should be assumed to be less than RD_{trial} (see Stegenga [2015], who argues that trials are systematically tuned to overestimate effectiveness). If $RD_{target} > 0$, then we have some evidence for an effectiveness hypothesis.

We use $RD_{patient}$ to represent the extent to which the intervention will change the probability that a particular patient will experience the outcome of interest. This is a prediction about the future outcome of a single case, and thus $RD_{patient}$ is a *subjective probability*. One makes an inference about $RD_{patient}$ based on one's knowledge of RD_{target} and the extent to which the patient is like the general population about which RD_{target} applies. As above, we do not think that $RD_{patient} = RD_{target}$. But the expected frequency in the general population, inferred from the actual frequency of the RCT, is at least a guide to the subjective probability that a particular patient will benefit from the intervention.

In sum, to estimate the probability that a medical intervention helps a particular patient, the chain of inferences is

$$RD_{trial} \longrightarrow RD_{target} \longrightarrow RD_{patient}$$

The first inference, from RD_{trial} to RD_{target}, is an extrapolation (or generalization)—an inference from a measurement performed in a study to an expectation about the frequency of people who will benefit from the intervention in a general target population. The second inference, from RD_{target} to $RD_{patient}$, is a particularization—an inference about the extent to which a particular person is similar in relevant respects to those people for whom the generalized inference of effectiveness was made (Fuller and Flores 2015).

For simplicity, we drop the subscripts and simply refer to outcome measures in the remainder of this chapter, though we specify the substantive interpretation of the outcome measure where necessary.

Frequency of Use of Various Outcome Measures

Researchers employ numerous outcome measures for reporting results produced by medical studies. These results are statistical in nature and relate to changes in risk associated with a control and an intervention group. The most popular outcome measures for dichotomous parameters are the five we introduced above: odds ratio, relative risk (or risk ratio), relative risk reduction, risk difference, and number needed to treat. The first three are standardly categorized as relative measures, and the last two are described as absolute measures. Relative measures are proportional measures quantifying, in terms of probabilities, the observed size of a treatment effect compared with an alternative intervention or control group. Absolute measures, in contrast, measure the observed size of a treatment effect, again expressed in probabilities, compared to a baseline risk. These different effect size measures convey different information. However, surveys of both systematic reviews of medical research and medical science journals demonstrate a marked preference among researchers for relative measures over absolute alternatives in reporting scientific results.

A review of research articles published in high-quality medical journals found that 150 out of 222 sampled RCTs and cohort studies neglected to communicate an absolute risk measure in the abstract of the study. Of the 150 studies that failed to report absolute measures in the abstract, 50 percent failed to report absolute risk at all (Schwartz et al. 2006).

In a more recent review, King, Harper, and Young (2012) found that out of 344 research articles published in quality journals in the field of health inequalities, only 138 reported a risk, measure in the abstract. Of these 138 studies, 122 reported only a relative risk, compared to 9 and 3 that reported an absolute risk only and both an absolute and relative risk, respectively.

In terms of reporting risk measures in the body of the research article, 258 of the 344 articles reported only relative measures, with only 139, or 53.9 percent, that included enough information to calculate an absolute outcome measure. In contrast, a mere 61 out of 344 reported only an absolute measure of effect in the body of the study, with less than half this number (25 out of 344) explicitly conveying both an absolute and a relative measure. And these results, the authors report, were consistent across journals, exposures, and outcomes (King, Harper, and Young 2012).

The disparities in reporting relative measures of effect size are even more pronounced in systematic reviews. Alonso-Coello et al. (2016) concluded an analysis of reporting frequencies of relative and absolute effect sizes within the title, abstract, and main body of 202 systematic reviews (94 Cochrane Collaboration reviews and 108 non-Cochrane reviews). The authors found that only 73 of 202 reviews reported absolute outcome measures for most patient-relevant parameters. Of these, only 12.3 percent advertised absolute outcome measures for both benefit and harm outcomes; harm outcomes, on the other hand, were reported entirely as absolute estimates. Exclusive reporting of beneficial outcomes in terms of relative risk occurred in approximately 80 percent of the systematic reviews (Alonso-Coello et al. 2016).

ABSOLUTE MEASURES ARE SUPERIOR TO RELATIVE MEASURES

We saw above that there are several commonly used outcome measures in medical research and that relative measures, such as RR, are more frequently used to report measurements than are absolute measures, such as RD. This is unfortunate because absolute measures are superior to relative measures. We defend this claim in this section.

The use of RR and RRR contributes to pervasive misinterpretations of data from medical research. Relative measures tend to make interventions appear more efficacious than they truly are. Consider again our example in Table 3.2. It is natural to take the reported RRR = .40 and infer that one will decrease one's overall probability of death by 40 percent if one opts for the treatment. Or consider another example: if prior epidemiological data indicate that patients have a 60 percent chance of dying from an infection and if an

RCT shows that a treatment decreased mortality in the test population by 40 percent, then there is a tendency for both medical care providers and patients to conclude that the chance of mortality after using the treatment reduces to 20 percent. In other terms, if 60 out of 100 people die from an infection, an advertised RRR = .40 conveys that the treatment has a capacity to decrease the mortality rate to 20 out of 100.

Unfortunately, relative measures tell us nothing about the probability of the outcome upon using a treatment. To find this quantity, we need to use the absolute outcome measure: risk difference (RD). RD tells us the change in overall probability of an outcome due to an intervention. Assuming the control group provides reliable information regarding the relevant base-rate probability of an outcome, RD measures the change in this probability. In our example in Table 3.2, this change in probability is from .20 to .12. The shift in risk measuring .08 is not as noteworthy as the larger RRR of .40 and does not make the intervention appear as impressive. This may go some way in explaining the prevalence of reporting RRR in pharmaceutical trials. Indeed, even manufacturers of pharmaceuticals have raised this issue. Here, for instance, is a passage from the Association of the British Pharmaceutical Industry (ABPI) code of practice:

> Referring only to relative risk, especially regarding risk reduction, can make a medicine appear more effective than it actually is. In order to assess the clinical impact of an outcome, the reader also needs to know the absolute risk involved. In that regard, relative risk should never be referred to without also referring to the absolute risk. Absolute risk can be referred to in isolation.

Unfortunately, as we saw in the section above, this guidance is not generally followed.

The ubiquitous practice of reporting solely RR and RRR (or emphasizing these relative measures while deemphasizing their absolute counterparts when reporting results) is especially problematic given the patient-care goal of clinical research. Ultimately, it is inferences about the extent to which a particular patient will benefit from an intervention that are necessary for making decisions about patient treatments. To determine this, one needs to estimate the probability that a patient will experience the outcome of interest if that patient uses a treatment minus the probability that a patient will experience the outcome of interest if that patient does not use that treatment. In the notation from the above section "Outcome Measures," one needs to estimate $RD_{patient}$. That is, one needs to estimate the extent to which the intervention will be a difference maker for this patient, and estimating that requires knowing the difference between the two conditional probabilities P(Y|E) and P(Y|C), which is just RD.

One must incorporate both of these conditional probabilities into a clinical decision problem, along with a specification of possible actions and the relative desirability of possible outcomes—for example, the seriousness of the outcomes and the monetary costs affiliated with treatments—to afford optimal patient-care decisions (here, we assume for simplicity standard decision theory, in which the decision principle is to maximize expected utility). The basic features of a decision context are the following:

- A range of possible actions, one of which must be chosen.
- Several exhaustive states of nature (outcomes), which may or may not obtain.
- An assignment of desirability or utility for each of the possible outcomes.
- The probabilities of the occurrence of the possible outcomes.

Consider, for example, a simplified decision problem in which a decision maker attempts to achieve some desirable end via one of two actions. Each action has associated with it the probability of achieving the desired end and either a cost or benefit depending on the desirability of the outcome. For illustrative purposes, we will describe the relative desirability of the outcomes in terms of utils, the standard unit of measurement from classical expected utility theory.

Suppose you face two possible actions: attend either the cricket match or the swim meet, but not both. The goal is to join a good friend at one of these two activities and avoid attending either event alone. Unfortunately, you neglected to prearrange the meet and otherwise have no ability to coordinate attendance with your friend in the interim. You know, however, that your friend will surely appear at one event, and mostly likely the swim meet.

To see the decision problem a bit more clearly, consider a simple two-by-two table with the associated probabilities of the outcomes in brackets, and the utils of each outcome in the cells of the table (see Table 3.3).

From Table 3.3, we see that if you go to the cricket match, there is a .40 probability that you will meet your friend there and a .60 probability that your friend will settle upon the swimming event instead. Conversely, if you attend the swim meet, there is a .60 probability that you will encounter your friend and a .40 probability that you will not. These are the conditional probabilities that specify the relative likelihoods of the possible state of affairs, namely, meeting your friend or not. Along with these probabilities, we have the utilities of each outcome: you receive 5 utils if you meet your friend at the cricket match, 5 utils if you meet at the swim meet, and 0 utils otherwise.

Table 3.3 Hypothetical decision scenario

Actions	*Outcomes*	
	Friend: Cricket (.40)	Friend: Swim (.60)
You: Cricket	5	0
You: Swim	0	5

According to expected utility theory, to combine the information offered by the scenario to make a rational decision, you must add the utilities for each action and multiply these sums by their associated probabilities. In other words, you must calculate the expected utility of each action and determine the action that maximizes expected value. Here the expected utilities for attending the cricket and swimming events are as follows:

- Cricket = (5 + 0) x .40 = 2
- Swim = (5 + 0) x .60 = 3

The action "go to the swim meet" maximizes expected utility.

In the absence of the conditional probabilities provisioned in the specification of the scenario, you have no rational basis on which to prefer one action over the other. Moreover, if the decision maker has no relevant conditional probabilities but finds herself in possession of new information that makes the friend's appearance at the cricket match more likely than it was previously, the decision maker still has no rational basis on which to prefer one action over the other. The probabilities of meeting your friend (or not) at an event are necessary for making the rational decision.

Translated into a patient-care decision context, these features of a decision context follow:

- The actions are treatment options: "undergo treatment" or "do not."
- The states of nature are the occurrence or nonoccurrence of a particular outcome.
- The utility assignment of the patient specifies his preferences regarding the costs and benefits associated with the occurrence and nonoccurrence of the outcome.
- The probabilities of the outcomes are the two conditional probabilities, P(Y|E)—the probability of the outcome given that the patient chooses the treatment—and P(Y|C)—the probability of the outcome given that the patient chooses not to receive the treatment.

Note that to satisfy the fourth feature of the patient-care decision context and avoid committing the base-rate fallacy, what is needed is $RD_{patient}$. As argued above, that is in part estimated by RD_{target}, which in turn is estimated by RD_{trial}. Knowing RR or other relative measures will not suffice. To see this, consider two studies of two drugs for two different diseases. The two drugs are x and y, and the two studies both have an RR of 0.7 for avoiding death. Suppose the costs and the side effects of the two drugs are the same and the only benefit the drugs bring is their reduction in one's risk of death. Suppose you have both diseases but you can choose only one of the drugs. Which should you choose? If you know only that both drugs have an RR of 0.7, you do not have enough information to choose. It would be a fallacy to suppose that they are equally good. To see this, suppose that in the first study, $P(\text{death}|x) = 0.07$ and $P(\text{death}|\text{no } x) = 0.1$. In the second study, $P(\text{death}|y) = 0.35$ and $P(\text{death}|\text{no } y) = 0.5$. In terms of absolute probabilities, drug x reduces one's probability of mortality by 0.03, and drug y reduces one's probability of mortality by 0.15. Knowing this, drug y is clearly superior, though to determine this, one had to compare the *difference* in the pertinent conditional probabilities; the *ratio* of the pertinent conditional probabilities was insufficient. That is, RD is necessary and sufficient to choose drug y over drug x, and RR is neither necessary nor sufficient.[1]

Neither patients nor medical care providers can make optimal treatment decisions with just relative outcome measures. Relative risks provide no information about how likely it is that patients will experience overall reductions (or increases) in risks when given an intervention. And as demonstrated above, it is this latter quantity that plays the important role in the patient-care decision context. The use of relative outcome measures in health care decision making, in fact, encourages health care providers and patients to commit the base-rate fallacy.

The base-rate fallacy occurs when one draws an inference regarding the overall probability of a hypothesis based on the comparative likelihoods of the evidence given the competing hypotheses without considering the relevant prior probabilities of those hypotheses (or "base rates"). The following example is often used to illustrate the base-rate fallacy.

Gabby receives a tuberculosis (TB) skin test performed at her local clinic. After the requisite wait time, Gabby finds out the skin test produced a positive indication (i.e., the test indicates that Gabby has TB). She discovers further that if she has TB, the test correctly identifies 9 out of 10 times and if she does not have TB, the test incorrectly identifies 1 out of 10 times. On this information, Gabby infers (incorrectly) that there is a .90 probability she has TB.

Gabby has committed the base-rate fallacy. She interpreted the likelihood of the positive result given that she has TB, namely, $P(+|TB)$, as the probability that she has TB given the positive result, that is, $P(TB|+)$. To calculate $P(TB|+)$, along with the likelihoods, Gabby requires the prior probability that she has TB—the prior probability of having TB is the "base rate" of the disease.

We may ascertain Gabby's prior probability of having TB from epidemiological data regarding the base rate of tuberculosis found within her reference class. That is to say, we require information on the relative frequency of TB cases within Gabby's reference class. Presuming she resides in the United States, the prior probability that Gabby has TB is fairly low, around .00003.

Before proceeding to the mathematical argument, intuitively we can see where Gabby's inferential mistake occurs. In drawing her conclusion, she fails to note that the TB test is many times more likely to produce a false-positive result than is a random selection of Americans to produce an individual with TB: P(+|No TB) = .10 >> P(Gabby has TB|Gabby is a randomly selected American) = .00003. In fact, we can see that for every randomly selected American who truly has TB, we can expect to observe approximately 333 false-positive TB test results; false-positive results are, in other words, 333 times more likely to occur than a true case of TB from a randomly chosen person in the United States.

Mathematically, we can show Gabby's error in more detail via Bayes' Theorem. This is a formal device used to calculate a conditional probability (the probability of one event occurring given that some other event(s) occurred). The conditional probability of interest in our present example is the probability that Gabby has TB, given that she tested positive for the infection on a reliable diagnostic test. Bayes' Theorem comes in several different, but mathematically equivalent, forms. Here is one version:

$$P(TB|+) = P(TB) \times P(+|TB) / ([P(TB) \times P(+|TB)] + [P(No\ TB) \times P(+|No\ TB)]),$$

where

- P(TB|+) = The probability that Gabby has TB given that she tested positive on a TB test.
- P(TB) = The probability that Gabby has TB.
- P(No TB) = The probability that Gabby does not have TB.
- P(+|TB) = The likelihood that the test produces a positive result given that Gabby has TB.
- P(+|No TB) = The likelihood that the test produces a positive result given that Gabby does not have TB.

The aim is to determine P(TB|+). To calculate P(TB|+), we need to provide the relevant probabilities for each term of Bayes' Theorem. We previously stipulated the likelihoods: P(+|TB) = .90 and P(+|No TB) = .10. And we can determine P(TB) from the epidemiological data mentioned above and deduce P(No TB) (via the complement rule for mutually exclusive events): P(TB) = .00003 and P(No TB) = 1 – P(TB) = 1 – .00003 = .99997.

With these values, we derive P(TB|+) as follows:

$$P(TB|+) = .00003 \times .90 / ([.00003 \times .90] + [.99997 \times .10])$$

$$P(TB|+) = .000027 / .100024$$

$$P(TB|+) = .00027$$

We see, then, the probability that Gabby has TB is .00027, not .90 as she inferred. The moral of the example: In both ignoring the antecedent improbability of her having TB and employing only the likelihoods to draw her conclusion, Gabby commits the base-rate fallacy. This fallacy, in turn, prevents Gabby from accurately estimating her true probability of having TB.

With the notion of the base-rate fallacy, we can articulate another virtue of RD compared with RR. A patient wants to know how the probability of having a particular outcome given that he receives the treatment compares with the probability of having a particular outcome given that he does not receive the treatment. What form should this comparison take: a *difference* between these conditional probabilities or a *ratio* of these conditional probabilities? The question is, *how* should we compare P(Y|E) and P(Y|C)? Since the outcome in question is Y, these conditional probabilities represent "posterior probabilities," so according to Bayes' Theorem, these conditional probabilities have the prior probability of the outcome, P(Y), "built in," so to speak. If we take the difference between the conditional probabilities, that difference continues to have the priors built in. If we take the ratio between the conditional probabilities, the prior, which is built in to both the numerator and the denominator, divides out. Thus, the *difference* between the conditional probabilities, which is RD, is determined in part by P(Y), but the *ratio* between the conditional probabilities, which is RR, is not determined at all by P(Y).

We will show this in less colloquial terms. By applying Bayes' Theorem, RR is equivalent to the following:

$$RR = [P(E|Y)P(Y)/P(E)] / [P(C|Y)P(Y)/P(C)]$$
$$= [P(E|Y)/P(E)] / [P(C|Y)/P(C)]$$

The baseline probability of having outcome Y, P(Y), has fallen out of the equation. Thus, RR is not sensitive to P(Y). In contrast, by applying Bayes' Theorem, RD is equivalent to the following:

$$RD = [P(E|Y)P(Y)/P(E)] - [P(C|Y)P(Y)/P(C)]$$

$$= P(Y)[[P(E|Y)/P(E)] - [P(C|Y)/P(C)]]$$

The leftmost term is the prior probability of Y. Thus, RD is sensitive to P(Y).[2] In other words, RD helps one avoid committing the base-rate fallacy, and RR facilitates the base-rate fallacy. Thus, RD is superior to RR.

THE MYTH OF THE ODDS RATIO

A widely held myth is that in case-control studies, only the odds ratios can be used as an outcome measure relating the rates of exposure among subjects with the disease (cases) and subjects without the disease (controls). This would be unfortunate since as we argued above, absolute measures (such as RD) are superior to relative measures (such as OR). Here, in more detail, is the myth:

> To compute relative and absolute risks from study populations, the study populations must constitute a population-based sample. That is to say, the study population must be a representative sample from the overall population of at-risk individuals. Case-control populations, however, are not a representative sample of the larger at-risk population. Epidemiologists deliberately select cases in case-control studies on the basis of inclusionary criteria, particularly the presence of a particular disease; similarly, epidemiologists deliberately select controls based on both matching criteria (i.e., criteria that ensure that controls are comparable to the cases in all relevant respects [except for the presence of the disease]) and the absence of the disease. Hence, case-control studies do not allow for the calculation of absolute or relative risks. However, case-control studies permit the calculation of sample odds ratios. Sample odds ratios estimate population odds ratios, which under the rare disease assumption, estimate the larger population relative risk. Thus, odds ratios are the appropriate outcome measure in case-control studies.

This myth is widely expressed in epidemiological literature.[3]

This myth is importantly related to the argument regarding the superiority of absolute measures, such as RD. If the myth is correct, then case-control studies do not allow us to compute the very outcome measure that above we argued was superior. However, the myth is not correct. We argue here that one can reasonably compute RR or RD from case-control studies.

In a case-control study, one starts with a set of people who have some outcome (call this set Y), and within this set one counts how many were exposed (call this subset E) and how many were not (C). The members of Y are “matched” with controls who do not have the outcome (call this set N), and again, within N one counts how many were exposed (E’) and not exposed

(C'). The numbers in E, C, E', and C' we can refer to as a, c, b, and d. These are numbers that we have access to in a case-control study.

Note that the common claim that case-control studies allow for the computation of neither absolute nor relative risk does not depend on a mathematical limitation. Consider that odds ratios are ratios of odds. Odds can be translated into probabilities. The odds of *X* is the probability of *X* divided by the probability of *not X*. Thus, for example, 3:1 odds in favor of *X* amounts to P(X) = .75 and P(not-X) = .25, and .75/.25 = 3/1. In case-control studies, the computed odds are conditional probabilities, but the conditional events are the probability that individuals were exposed given that they have a disease (or any other measured parameter), compared with the probability that individuals were exposed given that they do not have the disease. Now recall Table 3.1. A case-control study furnishes us with all of a, b, c, and d. That is, a case-control study furnishes us with all we need to calculate not just OR but also RR, RRR, RD, and NNT.

We saw above that RR and RD are based on the two conditional probabilities P(Y|E) and P(Y|C). Those are determined by the relative frequencies a/(a+b) and c/(c+d). *One can calculate these relative frequencies from the numbers generated in a case-control study.* It is true that one cannot calculate either [a/(a+b)] or [c/(c+d)] from either group of a case-control study alone because in the Y group one only has numbers a and c, which is obviously insufficient to compute [a/(a+b)], and vice versa for the N group. But taking the results from both arms of the case-control study together one has all the numbers one needs to compute RR because one has all of a, b, c, and d.

For illustration, suppose 50 people develop a disease. We include each of these people into our case-control study as cases and 150 individuals, comparably matched in every way save for the presence of the disease, as controls. We then review medical histories to determine relative binary exposure outcomes—exposure and nonexposure—for each population. If we note that of the 50 cases, 48 were exposed and of the 150 controls, 48 were exposed, then we have a sample consisting of 96 exposures and 104 nonexposures. From this sample, we calculate that the probability of developing the disease given exposure is 48/96 and the probability of not developing the disease given exposure is 48/96. Similarly, we calculate the probability of developing and not developing the disease given nonexposure as 2/104 and 102/104, respectively.

If we have ensured relative comparability between both controls and cases and have not otherwise populated our sample on grounds likely to bias exposure results, we may construe the case-control population as a randomly selected sample. It is as if we have drawn 200 marbles from a large bag of marbles, each of which is colored either red (for cases) or blue (for controls) and labeled with either a 1 (for exposure) or 0 (for nonexposure) to estimate the probabilities of association between color and labeling. We group our sample into 50 red marbles and 150 blue marbles and find 48 red marbles labeled with 1, 2 red marbles with 0, 102 blue marbles with 0, and 48 blue marbles with 1. The probabilities here remain identical to those above in the epidemiological example:

- The probability of drawing a red marble given it is labeled 1 = 48/96.
- The probability of drawing a blue marble given it is labeled 1 = 48/96.
- The probability of drawing a red marble given it is labeled 0 = 2/104.
- The probability of drawing a red marble given it is labeled 0 = 102/104.

In the marble example, however, we see much more easily the intuitiveness of calculating the conditional probability of drawing a marble of a particular color given it has a particular label. Think of an identical sample of marbles—same number of marbles and identical color and labeling sample distribution—only this time we first record the labeling and then note the color. In this latter scenario, nobody would find calculating, say, the probability of drawing a red marble given that it is labeled 1 as problematic. Indeed, if two different researchers held two different, but statistically identical samples, we would find a prohibition on the use of the sample relative frequencies to estimate the same probabilities rather strange. Especially strange once we are told that neither sample was selected in any way on grounds likely to bias the outcome under investigation. In short, in both the epidemiological and marble examples, insofar as we have ensured relative comparability between red marbles/cases and blue marbles/controls and have not otherwise populated our sample on grounds likely to bias exposure results, we have a random sample of observations.

Since no mathematical limitation exists to computing RD or other outcome measures from a case-control study, the myth must be based on something else. Perhaps the limitation is methodological. In a case-control study, the number of cases and controls is deliberately determined by the researcher, unlike the study populations in prospective studies, such as RCTs. The worry is that case-control samples are not representative of the larger at-risk population; put another way, case-control samples do not estimate the larger at-risk population.

If the case-control study population were representative, the claim goes, then we may compute either absolute or relative risks (or both) since we could infer that the observed association between exposure and disease results from the real causal relationship that occurs in the larger at-risk population.

The immediate difficulty with this line of reasoning is that no clinical sample, whether epidemiological or experimental, is representative in this respect. Populations of subjects in RCTs are not representative of the general population, and thus results of RCTs are not externally valid.[4]

The use of both inclusionary and exclusionary criteria in subject recruitment in RCTs ensures that study populations are not representative of target populations regarding a great number of relevant parameters. Hence, it is not clear why the unrepresentativeness of case-control populations should pose problems that the unrepresentativeness of subjects in RCTs do not.

In any case, it is not at all clear that a study population need be perfectly representative of the underlying population to determine absolute and relative risks. Although the number of cases and controls is predetermined by an epidemiologist, the rates of exposure are not. The rates of exposure among both cases and controls are, from the researcher's epistemic position, random variables. They are quantities distributed within a study and target population according to an unknown probabilistic distribution. One can use the sample data to draw inferences about this distribution.

Hence, even though the precise number of cases and controls is determined by researchers, given that the exposure rates are unknown, it is as if one mixed the study population in a bag and randomly assigned exposure or nonexposure to random draws from the bag. One can then calculate the probability that a subject developed the disease given that a subject was exposed. Whereupon, one may calculate both RR and RD, and not just OR. Since we argued earlier that absolute measures, such as RD, are superior to relative measures, such as RR or OR, medical scientists should use RD. Here, we have argued that despite the myth to the contrary, there is no principled reason why medical scientists cannot use RD for analyzing and reporting data from case-control studies.

CONCLUSION

In this chapter, we have argued that absolute measures, such as risk difference, are superior to relative measures, such as relative risk. Moreover, we argued that absolute measures, such as risk difference, are applicable in case-control studies despite the widespread myth that only (relative) odds ratios can be calculated from case-control studies.

NOTES

1. For a systematic decision-theoretic proof of the superiority of absolute measures, see Sprenger and Stegenga (forthcoming).
2. This proof first appeared in Stegenga (2015).

3. See Rothman, Lash, and Greenland (2008) and Vandenbroucke and Pearce (2012).
4. For a statement of this criticism of RCTs, see Stegenga (2015).

REFERENCES

Alonso-Coello, Pablo, et al. 2016. "Systematic Reviews Experience Major Limitations in Reporting Absolute Effects." *Journal of Clinical Epidemiology* 72: 16–26. doi: 10.1016/j.jclinepi.2015.11.002.

Fuller, Jonathan, and Luis J. Flores. 2015. "The Risk GP Model: The Standard Model of Prediction in Medicine." *Studies in History and Philosophy of Science Part C: Studies in History and Philosophy of Biological and Biomedical Sciences* 54: 49–61. doi: http://dx.doi.org/10.1016/j.shpsc.2015.06.006.

King, Nicholas B., Sam Harper, and Meredith E. Young. 2012. "Use of Relative and Absolute Effect Measures in Reporting Health Inequalities: Structured Review." *British Medical Journal* 345: e5774. doi: 10.1136/bmj.e5774.

Rothman, Kenneth J., Timothy L. Lash, and Sander Greenland. 2008. *Modern Epidemiology*. 3rd edition. Philadelphia: Lippincott Williams & Wilkins.

Schwartz, Lisa M., Steven Woloshin, Evan L. Dvorin, and H. Gilbert Welch. 2006. "Ratio Measures in Leading Medical Journals: Structured Review of Accessibility of Underlying Absolute Risks." *British Medical Journal* 333: 1248. doi: 10.1136/bmj.38985.564317.7C.

Sprenger, Jan, and Jacob Stegenga. Forthcoming. "Three Arguments for Absolute Outcome Measures." *Philosophy of Science*.

Stegenga, Jacob. 2015. "Measuring Effectiveness." *Studies in the History and Philosophy of Biological and Biomedical Sciences* 54: 62–71.

Vandenbroucke, Jan P., and Neil Pearce. 2012. "Case-Control Studies: Basic Concepts." *International Journal of Epidemiology* 41 (5): 1480–89. doi: 10.1093/ije/dys147.

Chapter 4

A Causal Construal of Heritability Estimates

Zinhle Mncube

What is the point of the statistical measure of heritability? Heritability estimates "have been regarded as important primarily on the expectation that they would furnish valuable information about the *causal* strength of genetic influence on phenotypic differences" (Sesardic 1993, 399; slightly rephrased in Sesardic 2005, 22). Height differs among people, for example. If height has a high heritability in a particular population, then the differences in height in that population are mostly due to genetic differences in the population rather than any environmental differences.

The purpose of this chapter is to ask the question, how might we causally interpret heritability claims? However, to ask this question, I must deal with the prior question—does it *ever* make sense to causally interpret heritability claims? The widely held answer to this question is no. Theorists use three main lines of argument to establish that heritability estimates are causally uninterpretable: (a) the existence of gene-environment interaction, (b) the existence of gene-environment correlation, and (c) the locality of heritability estimates.

In this chapter I argue that the view that "heritability estimates are devoid of causal implications" (Sesardic 2005, 10) is too quick. We know that genes have a causal effect on phenotypes; therefore, there is a prima facie reason to think that what heritability estimates measure is causal. I concentrate specifically on the existence of gene-environment interaction (of which there are two types—biometric and developmental). I show that it is possible to dissolve the challenges that the existence of gene-environment interaction creates for a causal construal of heritability estimates. Specifically, when there is no biometric gene-environment interaction (among other conditions),[1] it makes sense to causally interpret a heritability estimate as a measure of the causal strength of differences in genes on total phenotypic variance (Sesardic

2005; Tal 2009, 2012). Put differently, I propose that the challenges to a causal construal of heritability claims can be used to outline conditions under which heritability claims can be well justified and potentially generalizable, often depending on empirical matters that cannot be stipulated in advance.

I begin by explaining what heritability analysis is and why we should care about it. In the subsequent sections, I explain that there are two notions of gene-environment interaction: the biometric notion of gene-environment interaction, $G\times E_B$, and the developmental notion of gene-environment interaction, $G\times E_D$ (Tabery 2007a, 2008, 2014; Griffiths and Tabery 2008). I formalize the two different challenges that each sense of gene-environment interaction creates for a causal construal of heritability estimates, and I show how these challenges can be dissolved. Lastly, I outline some of the lessons we can take on causally construing heritability claims.

WHAT IS HERITABILITY?

Since the birth of behavior genetics as a discipline in the 1950s, heritability analysis has been used to understand the relation between heredity and one specific phenotype—behavior (Stoltenberg 1997, 90). It was surmised that heritability estimates were "important primarily on the expectation that they would furnish valuable information about the *causal* strength of genetic influence on phenotypic differences" (Sesardic 1993, 399; slightly rephrased in Sesardic 2005, 22).

A key assumption of heritability analysis is that of additivity, that is, that genetic and environmental influences act *separately* on phenotypes. Or formally, in generating heritability estimates, it is assumed that the total phenotypic variance (V_P) of a particular trait in a population can be partitioned into variance due to genetic variance (V_G) and environmental variance (V_E) such that:

$$V_P = V_G + V_E$$

Of course, total phenotypic variance is not just made up of genetic and environmental variance. It is also made up of measurement error (V_{Error}), gene-environment interaction (V_{GxE}) and gene-environment correlation (Cov_{GE}) (moreover, the two latter components are each made up of several components). As I will explain later, these components are complications to the measurement of heritability. Most simply, gene-environment interaction and gene-environment correlation are additional sources of variation in the total phenotypic variance sum. If they are present in a particular study, then heritability cannot be calculated because additivity no longer holds.

But first, it is important to note that there are two senses of heritability: narrow sense heritability (h^2) and broad sense heritability (H^2). I will not go into the specifics of the difference between these two senses of heritability. What is important is that when theorists discuss the issues that arise within the heritability I am interested in, most of them explicitly limit the discussion to broad-sense heritability (e.g., Sesardic 2005 and Kaplan 2006). Therefore, when I talk about "heritability," henceforth, I refer to heritability in the broad sense.

Broad-sense heritability (H^2) is calculated thus:

$$H^2 = V_G / V_P$$

It is this sense of heritability that is understood as a "measure of the proportion of the variance in a particular trait in a particular population that is attributed with genetic variation in that population" (Kaplan 2006, 56). Practically, heritability estimates are generated from the analysis of variance (ANOVA)—a statistical analysis that is used to quantitatively partition all the causes of phenotypic variation in a population. The heritability estimates that are generated from these methods lie between 0.0, or 0 percent (i.e., no genetic variation in a trait because all variation in that trait is due to the environment), and 1.0, or 100 percent (i.e., all difference in a trait is a result of genetic variation in a population).

I have just explained what heritability is and how it is estimated. But why should we care about heritability estimates? For one, today heritability estimates are widely used in human disease genetics, among other fields. As Tenesa and Haley (2013, 147) explain, "Estimates of heritability quantify how much of the variation in disease liability in a population can be attributed to genetic variation." Additionally, Vissher, Hill, and Wray (2008, 258–59) argue that heritability is "so enduring and useful" because, among other reasons, it is useful in understanding "the genetic component of risk to disease, independently of known environmental risk factors." More specifically, it is useful "in determining the efficiency of prediction of the genetic risk of disease" (Vissher et al. 2008, 258–59).

As I mentioned, despite this kind of use of heritability estimates, the widely held view among philosophers today is that heritability estimates do not indicate the causal strength of genes on phenotypic variance (Downes 2016; Oftedal 2005). I reject this view. Tabery's (2007a, 2008, 2014; Griffiths and Tabery 2008) analysis of gene-environment interaction illustrates that there are in fact two different notions of gene-environment interaction at play in the literature even though theorists do not distinguish between them (i.e., the biometric notion of gene-environment interaction, $G{\times}E_B$, and the developmental notion of gene-environment interaction, $G{\times}E_D$. Therefore, in what follows, my strategy to rebut the claim that heritability estimates cannot be causal is to (a) formalize the challenge to heritability analysis brought by each notion of gene-environment interaction and (b) outline how both challenges can be dissolved.

THE G×E_B CHALLENGE

The biometric notion of gene-environment interaction, or G×E_B, is the "statistical measure of the breakdown in additivity between genotypic and environmental sources of variation" (Tabery 2008, 728). What does this mean? I will use Cooper and Zubek's (1958) classic experiment on rats to explain G×E_B. Cooper and Zubek bred several groups of rats and exposed them to several Hebb-Williams (1946) maze tests where the rats had to make their way from the start of a maze to a food source at the end of the maze. The rats that were relatively good at running these mazes after several attempts were classed as "maze-bright," and the rats that were relatively bad were classed as "maze-dull" (Kaplan 2006, 60).

Cooper and Zubek wanted to see what would happen if these same rats were bred in impoverished or restricted environments (i.e., cages with no toys) versus enriched environments (i.e., cages with several toys). What they found, however, was that in the enriched environments, both types of rats "performed similarly well" (Kaplan 2006, 60). Conversely, in the impoverished or restricted environments, both types of rats "performed similarly poorly" (Kaplan 2006, 60).

If we have to account for the total phenotypic variation in the maze-running performance of Cooper and Zubek's rats, we cannot do so by just adding the main effects of V_G and V_E. Rather, "when different genetic groups respond differently to the same array of environments, the additivity between V_G and V_E breaks down, requiring an addition to the equation in the form of G×E [G×E_B]" (Tabery 2007a, 120). As such, the existence of G×E_B has implications for a causal construal of heritability estimates. One pertinent implication is that when G×E_B is present in a study, heritability can no longer be easily calculated since the variations from genotype and the environment "are no longer independent" as a heritability sum requires (Tabery 2007a, 5). It is argued that continuing with a calculation of heritability, in the face of such G×E_B, would be an inaccurate picture of the genetic influence on the differences in phenotype. In other words, the heritability estimate will be inflated because it will also include this G×E_B (i.e., it will also be confounded by the environmental element in G×E_B) (Layzer 1974, 1262).

Therefore, formally, in the presence of G×E_B in a study, I infer that the strongest challenge you will encounter against causally construing a heritability estimate from the study is as follows:[2]

- (B1): Heritability analysis is "based on the assumption of additivity" (Sesardic 2005, 49). This is the idea that differences in genotype and in environments act separately on total phenotypic variance.
- (B2): $G \times E_B$ complicates the calculation of heritability because it generates a separate source of variation in the total phenotypic variance sum (Tabery 2007a).
- (B3): If there is strong $G \times E_B$ in a study, then heritability cannot be easily calculated for a particular trait (Tabery 2007a).
- (B4): Nonadditivity is rampant in nature (e.g., Block and Dworkin 1974; Dick and Rose 2002; Layzer 1974; Lewontin 1974, 1975); $G \times E_B$ is pervasive in nature.[3]
- (B5): Therefore, heritability estimates cannot have a causal interpretation, let alone a useful one (Lewontin 1974; Northcott 2008),[4] or heritability is "meaningless in terms of its causal explanatory content" (Tal 2012, 234).

Notice that the $G \times E_B$ challenge hinges on the question of additivity (or step B4 in the above dialectic). If it is true that $G \times E_B$ is pervasive in nature, then this means that heritability estimates can never be causally interpreted as measures of the influence of genes on phenotype. In what follows, I present the additivity reply to this challenge. I argue for the following conclusion: since it is the pervasiveness of $G \times E_B$ that is used to establish the strong conclusion that heritability is "meaningless in terms of its causal explanatory content" (Tal 2012, 234), given that the amount of $G \times E_B$ present in a study is an empirical question, when there is no $G \times E_B$ in a study, a heritability estimate can be generated and causally interpreted.

The Additivity Reply

Like Jensen (1969), Levin (1997), Sesardic (2005), Oftedal (2005), and Tal (2009, 2012), I agree that we cannot answer the question of the pervasiveness of $G \times E_B$ and nonadditivity in human biology *in a nonempirical way*.

As Tal writes, "Ultimately, it is an empirical issue whether G×E [$G \times E_B$] interaction exists with respect to a particular target trait, the given distribution of genotypes and the range of environments" (Tal 2012). Thus, to the effect that theorists such as Gray (1992) and Sterelny and Griffiths (1999) argue that additivity is rare in nature based solely on Lewontin's (1974) argument, without any empirical evidence, I agree with Sesardic (2005) that this is incorrect. That is, to the effect that B4 in the $G \times E_B$ challenge to heritability analysis is defended in an aprioristic manner, it cannot be used to establish the strong conclusion that heritability is "meaningless in terms of its causal explanatory content" (Tal 2012, 234).

It can be decided only on a case-by-case basis whether $G \times E_B$ is present in a study and therefore whether heritability can be calculated for that trait. In other words, we can say of a situation only whether there is or is not $G{\times}E_B$ and therefore, whether we can or cannot calculate heritability in a meaningful way. As Tabery (2015, 4) writes, "In short, the long history of interaction research as well as the recent history of interaction research both remind us that evidential evaluations should be directed at individual results rather than at the entire interaction approach" (Tabery 2015, 4).

But Sesardic (2005) takes the argument further. He argues not only that additivity is an empirical question but also that, in fact, it is *nonadditivity* (or $G{\times}E_B$) that is rare in biology. How does Sesardic establish this claim of nonadditivity? Sesardic remarks that it is surprising just how little empirical evidence Lewontin (1974) invokes in his claim for the pervasiveness of nonadditivity. Clearly then, Sesardic will have to provide some empirical evidence of his own to establish the opposite claim, namely, that it is nonadditivity that *is* rare.

Sesardic cites seventeen authors (i.e., "people doing empirical research in the relevant fields") who argue that significant $G{\times}E_B$ is rarely found in actual experiments (Sesardic 2005, 68). To be sure, what Sesardic wants to establish is that significant statistical gene-environment interactions have rarely been found. Thus, what is at issue here for him is not simply the presence of $G{\times}E_B$ in a study, but rather *significant* $G{\times}E_B$.

I would like to introduce two issues here. First, I have said that any type of $G{\times}E_B$—weak or strong—is problematic for a causal construal of heritability estimates. On the other hand, Sesardic says that "it is only in the presence of *strong* G–E interactions that it becomes difficult to put a useful causal interpretation on heritability coefficients" (Sesardic 2005, 71). I surmise that for Sesardic, only strong $G{\times}E_B$ is problematic for heritability analysis because he, like Jensen (1969), thinks that small $G{\times}E_B$ can simply be eliminated in the total phenotypic variance sum through a process called a "transformation of scale" (to return to a situation of additivity). Without going into the specifics of a transformation of scale, one pertinent methodological problem is that such a transformation of scale cannot be done without a justification. As Tabery (2007a, 5) writes, "If it is employed purely for the sake of statistical convenience without regard to any plausible biological framework, then it is unclear what biological information the measurement provides after the transformation has been performed".

Unless well motivated, attempting to eliminate $G{\times}E_B$ through a transformation of scale is problematic for a causal construal of the heritability estimate (Wahlsten 1990). This is part of the reason why I argue that it is only when there is no $G{\times}E_B$ (among other conditions) that we can causally construe heritability estimates. When there is no $G{\times}E_B$, there is obviously additivity, and more importantly, there can be no confounding of the heritability estimate with environmental or G×E_B effects. So the first issue is that Sesardic should be concerned with the presence of all types of G×E_B—small and strong.

Second, there is the methodological issue that the ANOVA cannot detect some G×E_B effects (Tal 2012; Wahlsten 1990). There is also "a theoretical issue whether the presence of undetected interaction renders partitioning of the phenotypic variance and heritability estimates invalid" (Tal 2012, 234). The question would then be that if I argue that heritability estimates can be causally construed when there is no G×E_B, given these issues, then could we ever trust a finding of no G×B_B?

It is beyond the scope of this chapter to survey and settle all these kinds of criticisms against ANOVA. I will say that ANOVA is thought to better detect G×E_B when the *sample size is large* (Wahlsten 1990). This is to say that not all heritability estimates will have causal import. For example, besides the condition of no G×E_B (when the sample size is large), there also needs to be little to no gene-environment correlation[5] in the study (Tal 2009, 2012). This is to say that it is more appropriate to use ANOVA in some contexts and not in others (Sesardic 2005, 86). Moreover, not all heritability estimates warrant a causal interpretation. Thus, it is true that unless it is specified that the sample size is large and there has been effort to find G×E_B, we must be careful about the type of causal interpretation we draw from heritability estimates.

As such, since the question of statistical gene-environment interaction in a study is an "open empirical question" (Sesardic 2005, 49), it cannot be argued from the existence of G×E_B that heritability estimates never warrant a causal interpretation. When there is no G×E_B in a study (among other conditions, such as minimal gene-environment correlation), it makes sense to interpret the heritability estimate as a measure of the causal relation between genetic variance and total phenotypic variance of a trait. In fact, Tal (2009, 2012) proposes a probabilistic interpretation of heritability estimates under such conditions. Therefore, the G×E_B challenge is dissolved.

THE G×E_D CHALLENGE

The developmental notion of gene-environment interaction, or G×E_D, is also problematic for heritability analysis. Whereas G×E_B has to do with explaining "the differences between individuals," G×E_D has to do with "how those individuals came to have the phenotypes that they do" (Griffiths and Tabery 2008). Or as Moore (2015, 418) puts it, even when the ANOVA does not find *statistical* G×E_B between genotype and environment, it "does not mean the phenotype develops in the absence of *mechanical* interactions [G×E_D] between DNA segments and their contexts."

This is the challenge that $G{\times}E_D$ creates for a causal construal of heritability:[6]

- (D1): $G{\times}E_D$ is "the causal-mechanical interaction between genes and the environment during individual development" (Griffiths and Tabery 2008). It is an intrinsic and pervasive part of the development of phenotypes.
- (D2) Even when there is no $G{\times}E_B$, it "does not mean that the phenotype develops in the absence of *mechanical* interactions between DNA segments and their contexts" (Moore 2015, 418).
- (D3): An attempt to measure heritability—even when there is no $G{\times}E_B$—ignores the complexity and interdependence "of the developmental genotype-environment-phenotype relationship" (Tabery 2007a, 91).
- (D4) "The *real* focus for geneticists . . . should be the causal mechanics of the developmental genotype-environment-phenotype relationship (Tabery 2007a, 67).
- (D5) Heritability cannot tell us about the causal mechanisms[7] behind the individual development of phenotypes (Daniels 1974, 170).
- (D6): Therefore, heritability estimates cannot warrant a causal interpretation, let alone a useful one, or heritability is "meaningless in terms of its causal explanatory content" (Tal 2012, 234).

Essentially, the $G{\times}E_D$ challenge is about how *little* ANOVA and the heritability estimates derived from it tell us about individual development. If we agree that there is gene-environment interaction at the level of individual development ($G{\times}E_D$), we can already see why theorists such as Lewontin (1974) would argue that partitioning total phenotypic variation into its alleged parts and then calculating heritability is ill formed. The $G{\times}E_D$ challenge is about how the ANOVA ignores causal mechanisms. Also, notice that the challenge assumes that only knowledge of the causal mechanisms behind individual development is sufficient for causal inference.

In what follows, I use Tabery's interdependent-difference-makers concept of gene-environment interaction[8] (or IDM_{GXE}) to reply to this $G{\times}E_D$ challenge. My reply deals primarily with steps D3 and D5 in the $G{\times}E_D$ challenge. My reply is this: the ANOVA does not ignore individual development. In fact, a high heritability estimate in a study could point to a genetic mechanism underlying the development of the trait (Sesardic 1993, 405; Sesardic 2005, 25).

The Interdependent-Difference-Makers Reply

What is the IDM_{GXE}? In this chapter, I have used Tabery's (2015) analysis to the effect that the long-standing debate about gene-environment interaction and its implications for heritability analysis is just one instance of the debate "between the variation-partitioning [or quantitative behavioral genetics][9] and the mechanism-elucidation [or developmental psychobiology][10] approaches" on the etiology of traits. Specifically, Tabery argues that each sense of gene-environment interaction originates from each competing approach—$G{\times}E_B$ from the variation-partitioning approach and $G{\times}E_D$ from the mechanism-elucidation approach. Tabery's analysis is helpful because it shows how, historically, opinions about the implications of gene-environment interaction for heritability analysis have differed depending on the notion of G×E employed by the theorist (see Tabery 2007a, 2008, 2014; Griffiths and Tabery 2008 for further detail).

However, Tabery (2007a, 138) believes that the two approaches are not "isolated as different levels of analysis" but should instead be integrated under the idea of difference mechanisms;[11] thus:

> In short, the difference-making variables in the mechanisms [genotypes and the environment] simultaneously are the causes of variation when the difference-making variables take different values in the natural world. . . . Individual differences, then, are the effect of the difference-makers in individual development when the difference-makers naturally take different values." (Tabery 2007a, 134).

Tabery then argues that this integration of these two approaches under difference mechanisms can also illuminate the debates around which notion of gene-environment interaction ($G{\times}E_B$ or $G{\times}E_D$) is correct. Tabery argues that we need not choose between $G{\times}E_B$ and $G{\times}E_D$. Instead, he proposes that both concepts be "integrated under the interdependent-difference-makers concept" of gene-environment interaction (Tabery 2007a, 148): "In the terminology of $G{\times}E_B$ and $G{\times}E_D$: $G{\times}E_B$ is a statistical measure of $G{\times}E_D$, which can itself be understood in more general, causal-mechanical terms as the result of the interdependence of difference-makers that take different values in the natural world" (Tabery 2007a, 174).

In my reply to the $G{\times}E_D$, I take my lead from the "$G{\times}E_B$ is a statistical measure of $G{\times}E_D$" in Tabery's (2007a, 174) explanation of the IDM_{GXE}. If an ANOVA happens to detect $G{\times}E_B$ in a study, I interpret this $G{\times}E_B$ to simply be a measure or reflection of "the interdependence of difference-makers in development [$G{\times}E_D$] *that take different values in the natural world*" (Tabery 2007a, italics added).

What the IDM_{GXE} shows is that when there is $G{\times}E_B$ in a study, "the difference-making variables [genotypes and the environment] in the mechanisms" (Tabery 2007a, 134) were (a) interdependent, (b) *actual* difference-makers, that (c) took "different values in the natural world" (Tabery 2007a, 174). In this instance, we should not measure heritability. If we did measure heritability, or if we attempted to eliminate $G{\times}E_B$, that heritability estimate would not warrant a causal interpretation as a measure of the causal relation between genetic variance and total phenotypic variance of a trait. This inability to form a causal interpretation is because the genetic variance would be confounded by the influence of $G{\times}E_B$.

But when there is no $G{\times}E_B$, we can measure heritability, and heritability estimates can be causally interpreted. This is possible because even if we agree that $G{\times}E_D$ might be significant in a study, that it did not reflect as significant $G{\times}E_B$ might mean that "the difference-making variables [genotypes and the environment] in the mechanisms" (Tabery 2007a, 134) *were* (a) interdependent, but they were not the (b) *actual* difference-makers, and they did not (c) take "different values in the natural world" (Tabery 2007a, 174).

This is the sense in which Tabery (and he adds, Douglas Wahlsten) believes that the ANOVA is "at its best" when it is detecting interaction effects ($G{\times}E_B$) and bad when it is trying to eliminate them (Tabery 2007a, 162). The detection of $G{\times}E_B$ is helpful "precisely because of the insights given for understanding development when an interaction effect is found" (Tabery 2007a, 172–73).

What are these "insights"? Clearly, from his use of difference mechanisms and the IDM_{GXE}, that for Tabery, the value of the studies in the variation-partitioning approach (i.e., heritability analysis) lies in that they can help us develop hypotheses about what the difference-making variables in the individual development of phenotypes *might* be (Moore 2015). Scientists in the mechanism-elucidation approach can then use these hypotheses to "elucidate the mechanisms that give rise to the phenotypes" (Moore 2015, 418).

The idea that the variation-partitioning approach can help to steer the mechanism-elucidation approach to hypotheses worth exploring, is like Sesardic's (1993) and Jensen's (1972) ideas about the use of heritability estimates generated from the ANOVA. Jensen claims that "a heritability study may be regarded as a Geiger counter with which one scans the territory in order to find the spot one can most profitably begin to dig for ore" (Jensen 1972, 243). Sesardic (1993, 405; Sesardic 2005, 24–25) adds here that we must understand heritability and the ANOVA as a first step in an etiological investigation. He (1993, 405; 2005, 25) writes, "High heritability of a trait (in a given population) often signals that it may be worthwhile to dig further, in the sense that an important genetic mechanism controlling differences in this trait may thus be uncovered."

As such, when I suggest causally interpreting a heritability estimate when there is no $G{\times}E_B$ (among other conditions), I am not ignoring individual development (*contra* D3 in the $G{\times}E_D$ challenge). No $G{\times}E_B$ in a study might mean that the interdependent "difference-makers in development" (or $G{\times}E_D$) have simply not taken "different values in the natural world" (Tabery 2007a,

174). In fact, a high heritability estimate could point to a genetic mechanism underlying the development of the trait (Sesardic 1993, 405; Sesardic 2005, 25) (*contra* D5 in the $G{\times}E_D$ challenge). Viewed in this light, heritability estimates can be causally interpreted as a measure of the strength of differences in genes on total phenotypic variance when there is no $G{\times}E_B$ even if there might be a $G{\times}E_D$ challenge. The $G{\times}E_D$ challenge is dissolved.

An Evaluation of the Reply

Let me now deal with some possible points of contention with my proposed solution. The first point of contention could be my choice of Tabery's IDM_{GXE} as a reply to the $G{\times}E_D$ challenge, and indeed the general integration of the VP-ME approaches that Tabery champions. Sesardic (2015, 1126), for example, questions Tabery's use of mechanisms and actual difference-makers to bridge the gap between the two approaches. Specifically, he insists that "[s]cientists cross this kind of 'explanatory divide' in the same way people cross the equator: without noticing that it exists" (Sesardic 2015, 1127).

To this I would say that if theorists on either side of the VP-ME bridge found value in the other's approaches there would, for example, be no talk of "muddle-headed" interaction (as when Sesardic [2005] refers to $G{\times}E_D$) or calls for heritability to be "given an honorable pension" (Rose 2006, 526). I used Tabery's framework and concepts because they show that scientists in the variation-partitioning approach (of which the ANOVA and heritability analysis are part) have not been engaged in some useless and invalid knowledge-gaining exercise. Moreover, Tabery's framework also highlights Sesardic's (2015) point that the final goal in the etiology of traits was always the elucidation of the underlying causal mechanisms.

The second point of contention could be the suggestion that the variation-partitioning approach is a possible first step to "elucidate the mechanisms that give rise to the phenotypes" (Moore 2015, 416). According to Moore, the integration of the VP-ME approaches propagated by Tabery creates the idea that these approaches are *equal* tools in the elucidation of the etiology of traits. But for Moore (2015, 416), "the tools on either side of Tabery's explanatory bridge are not of equal value." Moore insists that the variation-partitioning approach can give only a partial story about the etiology of traits because it cannot on its own add to our understanding of causal mechanisms. He insists that the variation-partitioning approach cannot, for example, "offer any practical information about how to influence the development of children, crops, or livestock in beneficial ways" (Moore 2015, 416).

For Moore, the mechanism-elucidation approach, on the other hand, is much more "self-contained" (Moore 2015, 416). He argues that "the mechanism-elucidation approach can identify tools that can be used to intervene in development in beneficial ways *and* can identify actual difference-makers that can help explain variation across a population" (Moore 2015, 416). This belies the question of why the ANOVA or heritability analysis, for example, should be a first step in identifying difference-makers in development, when the mechanism-elucidation approach can work all that out on its side of the bridge?

Notice Moore's assumption here that (a) only an elucidation of causal mechanisms can give a full story of the etiology of traits and (b) only an elucidation of causal mechanisms suffices for useful causal inference. About (a), Jensen's Geiger pointer analogy illustrates the point that the function of heritability analysis and the ANOVA was always to provide *some* of the causal story of the etiology of traits. As Sesardic (2015, 1127) contends, scientists in the variation-partitioning approach never did ignore or deny the importance of the elucidation of mechanisms. He writes, "They would have loved to know more about mechanisms but they realized that the goal was not achievable at the time. . . . Only after the recent explosion of knowledge about genes and brains could variation partitioning be complemented with a hunt for specific causal mechanisms."

About (b), first, even though we have more knowledge about causal mechanisms (Sesardic 2015, 1127) today in biology than at any other point (e.g., pioneering work by Machamer, Darden, and Craver [2000]), our mechanistic understanding (of phenotypes) is still limited (Aalen and Frigessi 2007, 159). To reiterate the Geiger counter analogy, Aalen and Frigessi explain that it is precisely in those areas where such mechanistic understanding is lacking that statistics (after all, heritability is a statistical measure) play their greatest, albeit modest, role (Aalen and Frigessi 2007, 159). Second, useful causal inferences about smoking and lung cancer, for example, were made some time before we had any (indeed we still do not have full) mechanistic understanding of the causal link between the two processes (Broadbent 2013, 77).

LESSONS ON HERITABILITY

A concern in the reader's mind about my argument might be that we might more easily or more reliably be able to measure heritability if there were no $G \times E_B$. However, this does not mean that what heritability measures is *causal.* It might be argued that all that heritability measures is a *mere statistical association* between genetic variance in a population and total phenotypic variation. Seen in this way, there is no need to think about a causal interpretation of heritability because at best, we do not know what that interpretation is or at worst, there is no causal interpretation of heritability.

Besides the idea that there is a prima facie reason to think that what heritability estimates measure is causal, I have two things to say here. First, heritability estimates are often described using causal language. For example, Block and Dworkin (1974, 51) explain: "The heritability of a characteristic tells us the degree to which variation in the characteristic is caused by genetic differences." Stoltenberg (1997, 90) suggests that we avoid terms like "'due to' or 'caused by' when referring to the statistical relations between an independent variable and a dependent variable (e.g., in an analysis of variance), but instead use terms such as 'associated with' to avoid deterministic implications."

The last part of the quote from Stoltenberg is important. He suggests avoiding causal language in this case not necessarily because there is absolutely no instance in which a heritability estimate would be legitimately describing a causal relationship between differences in genes and total phenotypic variance but rather because of how this relationship has been *misinterpreted* as implying some kind of deterministic implications.

Second, note the often-cited misuses of heritability estimates. Heritability estimates are sometimes misinterpreted as indicating how much a trait is *caused by* genes. So if the disease liability of diabetes is found to have a heritability of 1 in a particular population, this is sometimes misinterpreted as indicating that this disease liability is 100 percent caused by genes. But mathematically, a heritability estimate of 1 is clear (i.e., all the phenotypic variance in disease liability in diabetes in this population is attributable to the differences in genes in the population).

We know that heritability does not measure anything to do with the development of phenotypes, but rather differences in phenotypes. We also know that phenotypes cannot develop without the presence of both genes and the environment. The problem comes in when "attributable to" is *causally interpreted* as "completely caused by." Put differently, the biggest problems for heritability estimates arise not in their mathematical estimation but rather in their incorrect (as in the disease liability example) or inappropriate (when there is $G{\times}E_B$) *causal* interpretations.

Some further concerns might be the following: Is the use of heritability estimates not put into question if they can be causally interpreted *only* when there is no $G{\times}E_B$? Yes, heritability estimates offer a limited and very specific type of causal knowledge. Is what happens in development (as the existence of $G{\times}E_D$ shows) not so complicated that heritability estimates underplay the complexity of the causal story of phenotypes (Sesardic 2005, 87)? Of course, partitioning total phenotypic variance into its parts involves a simplification (Sesardic 2005, 87). This simplification is nonetheless causal and is nonetheless useful in a limited and specific way. For example, heritability estimates would be more useful if they told us not just that genetic mechanisms might be more highly involved in the trait differences but also how genes contribute to the development of traits. But they do not. Will the estimation of heritability become less desirable as we develop more empirical knowledge about the direct genetic mechanisms underlying phenotypes? I imagine so.

CONCLUSION

I have argued that it is possible to reply to one of the main lines of argument used to establish that heritability estimates are causally uninterpretable, namely, gene-environment interaction. Therefore, the consensus that "heritability estimates are devoid of causal implications" (Sesardic 2005, 10) is too quick. Specifically, I have argued heritability estimates can bear a causal interpretation when there is no statistical gene-environment interaction (Sesardic 2005; Tal 2009, 2012). When these conditions are met, it makes sense to causally interpret that heritability estimate as a measure of the causal strength of differences in genes on total phenotypic variance.

NOTES

1. These other conditions include (a) when there is little to no gene-environment correlation (Tal, 2009, 2012) and (b) only within the domain of populations that have similar causally salient features. They follow from the other main lines of argument against a causal construal of heritability estimates—the existence of gene-environment correlation and the locality of heritability estimates.

2. From the way that he argues, clearly this is the type of challenge that Sesardic is responding to in his 2005 book (additionally, Jensen [1969] and Levin [1997] were also responding to this kind of challenge against heritability).

3. One might question whether these two statements are as commensurate as I have put them here. As Block and Dworkin (1974, 54) touch on, "the claim that there is no genotype-environment interaction comes to much the same thing as the claim that IQ [or any phenotype] is a linear function of genotype and environment." I take this to mean that when we say there is no G×EB, we mean that the assumption of additivity is warranted. Conversely, when we say that there is pervasive G×EB, we mean that the assumption of *nonadditivity* is warranted.

4. Strictly, Lewontin (1974) and Northcott (2008) direct their arguments to the inability of the ANOVA to measure causal strength. However, heritability estimates are generated from ANOVA, and as Northcott (2008, 53) says, this criticism extends to heritability estimates.

5. There is gene-environment correlation in a study when "an individual's genotype correlates with exposure to particular environments" (Tabery 2007a, 40). This is problematic for a causal construal of the heritability estimates because it makes it

"impossible to know how much of the phenotype similarity arises from similarity of genotype and how much from the similarity of environment" (Feldman and Lewontin 1975, 1164).

6. This challenge is inferred from the arguments in Lewontin (1974), Layzer (1974), and Feldman and Lewontin (1975), among others.

7. A causal mechanism is "[a] step-by-step explanation of the mode of operation of a causal process that gives rise to a phenomenon of interest" (Nicholson 2012, 153).

8. I will not go into all the specifics of Tabery's solution here. I will just touch on the main points to explain my reply. For a more thorough analysis, see Tabery (2007a).

9. This is a "program devoted to measuring the relative contributions of nature and nurture to individual differences in populations" (Tabery 2007a, 26).

10. This is an approach that promises to "elucidate the causal mechanisms of behavioral development rather than quantify differences in behavioral development, and must ask *how* genes cause development rather than *how much* development genes cause" (Griffiths and Tabery 2008).

11. "Difference mechanisms are regular causal mechanisms made up of difference-making variables that take different values in the natural world" (Tabery 2007a, 130).

REFERENCES

Aalen, Odd O., and Frigessi, Arnoldo. 2007. "What Can Statistics Contribute to a Causal Understanding?" *Scandinavian Journal of Statistics* 34: 155–68.

Block, N. J., and Dworkin, Gerald. 1974. "IQ, Heritability and Inequality." Part 2. *Philosophy and Public Affairs* 4: 40–99.

Broadbent, Alexander. 2013. *Philosophy of Epidemiology*. London: Palgrave Macmillan.

Cooper, Roderick M., and John P. Zubek. 1958. "Effects of Enriched and Restricted Early Environments on the Learning Ability of Bright and Dull Rats." *Canadian Journal of Psychology* 12: 159–64.

Daniels, Norman. 1974. "IQ, Heritability and Human Nature." *PSA: Proceedings of the Biennial Meeting of the Philosophy of Science Association*, 143–80.

Dick, Danielle M., and Richard J. Rose. 2002. "Behavior Genetics: What's New? What's Next?" *Current Directions in Psychological Science* 11: 70–74.

Downes, Stephen M. 2016. "Heritability." http://plato.stanford.edu/entries/heredity/; accessed April 10, 2016.

Feldman, Marcus W., and Richard C. Lewontin. 1975. "The Heritability Hang-Up." *Science* 190: 1163–68.

Gray, Russell D. 1992. "Death of the Gene: Developmental Systems Strike Back." In *Trees of Life: Essays in Philosophy of Biology*, edited by Paul E. Griffiths. Dordrecht: Kluwer.

Griffiths, Paul E., and James G. Tabery. 2008. "Behavioral Genetics and Development: Historical and Conceptual Causes of Controversy." *New Ideas in Psychology* 26: 332–52.

Hebb, Donald O., and Kenneth K. Williams. 1946. "A Method of Rating Animal Intelligence." *Journal of General Psychology* 34: 59–65.

Jensen, Arthur R. 1969. "How Much Can We Boost IQ and Scholastic Achievement?" *Harvard Educational Review* 39: 1–123.

Jensen, Arthur R. 1972. "The IQ Controversy: A Reply to Layzer." *Cognition* 1: 427–52.

Kaplan, Jonathan M. 2006. "Misinformation, Misrepresentation, and Misuse of Human Behavioural Genetics Research." *Law and Contemporary Problems* 69: 47–80.

Layzer, David. 1974. "Heritability Analyses of IQ Scores: Science or Numerology?" *Science*, 183: 1259–66.

Levin, Michael. 1997. *Why Race Matters: Race Differences and What They Mean.* Westport, CT: Praeger.

Lewontin, Richard C. 1974. "The Analysis of Variance and the Analysis of Causes." *American Journal of Human Genetics* 26: 400–11.

Lewontin, Richard C. 1975. "Genetic Aspects of Intelligence." *Annual Review of Genetics* 9: 387–405.

Machamer, Peter K., Lindley Darden, and Carl F. Craver. 2000. "Thinking About Mechanisms." *Philosophy of Science* 67: 1–25.

Moore, David S. 2015. "The Asymmetrical Bridge: A Review of James Tabery's Book *Beyond Versus*." *Acta Biotheoretica* 63: 413–27.

Nicholson, Daniel, J. 2012. "The Concept of Mechanism in Biology." *Studies in History and Philosophy of Biological and Biomedical Sciences* 43: 152–63.

Northcott, Robert. 2008. "Can ANOVA Measure Causal Strength?" *Quarterly Review of Biology* 83: 47–55.

Oftedal, Gry. 2005. "Heritability and Genetic Causation." *Philosophy of Science* 72: 699–709.

Rose, Steven P. R. 2006. "Commentary: Heritability Estimates: Long Past Their Sell-By Date." *International Journal of Epidemiology* 35: 525–27.

Sesardic, Neven. 1993. "Heritability and Causation." *Philosophy of Science* 60: 396–418.

Sesardic, Neven. 2005. *Making Sense of Heritability.* New York: Cambridge University Press.

Sesardic, Neven. 2015. "Crossing the 'Explanatory Divide': A Bridge to Nowhere?" *International Journal of Epidemiology* 44: 1124–27.

Sterelny, Kim, and Paul E. Griffiths. 1999. *Sex and Death: An Introduction to Philosophy of Biology*. Chicago: University of Chicago Press.

Stoltenberg, Scott F. 1997. "Coming to Terms with Heritability." *Genetica* 99: 89–96.

Tabery, James G. 2007a. "Causation in the Nature-Nurture Debate: The Case of Genotype-Environment Interaction." (Doctoral dissertation). University of Pittsburgh.

Tabery, James G. 2008. "R. A. Fisher, Lancelot Hogben, and the Origin(s) of Gene-Environment Interaction." *Journal of the History of Biology,* 41: 717–61.

Tabery, James G. 2014. *Beyond Versus: The Struggle to Understand the Interaction of Nature and Nurture*. Cambridge, MA: MIT Press.

Tabery, James G. 2015. "Author's Reply: Considerations of Context, Distractions by Politics and Evaluations of Evidence." *International Journal of Epidemiology* 44: 1132–36.

Tal, Omri. 2009. "From Heritability to Probability." *Biology and Philosophy* 24: 81–105.

Tal, Omri. 2012. "The Impact of Gene-Environment Interaction and Correlation on the Interpretation of Heritability." *Acta Biotheoretica* 60: 225–37.

Tenesa, Albert, and Chris S. Haley. 2013. "The Heritability of Human Disease: Estimation, Uses and Abuses." *Nature Reviews Genetics* 14: 139–49.

Vissher, Peter M., William G. Hill, and Naomi R. Wray. 2008. "Heritability in the Genomics Era: Concepts and Misconceptions." *Nature Reviews Genetics* 9: 255–66.

Wahlsten, Douglas. 1990. "Insensitivity of the Analysis of Variance to Heredity-Environment Interaction." *Behavioral and Brain Sciences* 13: 109–61.

Part II

MEASURING INSTRUMENTS

Chapter 5

A Theory of Measurement

Norman M. Bradburn, Nancy L. Cartwright, and Jonathan Fuller

This chapter discusses basic issues about the nature of measurement for concepts in the social sciences and medicine and introduces a three-stage theory of measurement. In science and policy investigations we study quantities or concepts and their relations to understand and predict the behavior of individuals/tokens displaying those quantities or falling under those concepts. What does it mean to measure a quantity (e.g., body size) or to assign a concept or category (e.g., "underweight") to a token? In medicine, as throughout natural and social science, measurement is not just assigning categories or numbers; it is assigning values in a systematic and grounded way. This involves applying well-grounded metrics representing the quantity (e.g., body mass index [BMI]) to the token. This requires that:

1. We define the concept or quantity, which includes identifying its boundaries and fixing which features belong to it and which do not (*characterization*).
2. We define a metrical system that appropriately represents the quantity or concept (*representation*).
3. We formulate rules for applying the metrical system to tokens to produce the measurement results (*procedures*).

The reasons we undertake a measurement project—what we want to use the measurement results for—may affect one or more of these steps. Although 1–3 are listed as separate steps to help analyze measurement processes, what happens in each stage should influence other stages. We may, for example, come to recharacterize a category based on results derived relative to a candidate metrical representation of it. This may be the case with characterizations of *quality of life*, as discussed below. Or we may pick a metrical system because the procedural rules for applying it are well defined, or users know these methods better, or the methods are easier to implement.

All three steps—characterization, representation, and procedures—need explication. For an adequate measurement, these three must line up properly together: the representation of the quantity or quality measured must be appropriate to the central features taken to characterize it; equally, the procedures adopted to carry out the measurement must be appropriate to the formal representation adopted, and we should have good reason to expect that within acceptable bounds of accuracy the values assigned indicate the underlying values we aim toward. "Validating" a measure requires showing that it satisfies these requirements. This can be done formally, explicitly, and successfully. But these steps should not be neglected.

Broadly speaking, there are two metaphysical attitudes toward measurement concepts: realism and nominalism. These attitudes are analogous to (and overlapping with) realist and antirealist positions, respectively, regarding theoretical entities in the philosophy of science. Does a theoretical term such as "boson" or "prion" or a concept such as "hospital quality" or "mild cognitive impairment" refer to real features, or is the concept a *mere* name, a word used in an orderly way in language but not referring to anything in the world? If a measurement concept refers to real features, then "representing the concept" (using a table of indicators for "hospital quality" or a particular cognitive assessment scale for "mild cognitive impairment") can be taken to mean representing the features to which the concept refers. If, on the other hand, the concept does not refer to any real features, then representing the concept might mean something like "representing its internal structure." Finally, if a concept and its representation correspond to real-world features, then measurement procedures seek to assign values to tokens corresponding to the level or magnitude of those features that the token displays. But if the concept and its representation do *not* refer to real features, measurement procedures may simply seek to output values that can be used in inference and that satisfy certain desiderata, such as predictive success or reliable ordering of members of a set. We remain neutral as to whether the measurement concepts we will discuss here are real. The three-step account of measurement we present is consistent with both realist and nominalist perspectives.

THREE STEPS

Characterizing the Concept

The characterization of concepts in the sciences may be precise. The more precise the characterization, the more likely the concept is to be defined in terms of its measurement procedures (operationalization). But most concepts in social science, public health, and medicine—particularly in the policy realm—are loosely defined at the start. Further, the definitions may vary depending on the use to which the concept is put. For example, "disability" may mean different things depending on whether we are talking about an individual, a policy goal, a variable in a psychological theory, or a characteristic of a group of individuals. Concepts used in political discussions are often used in loose senses.

Many social science and medical concepts seem to refer to specific qualitative features (e.g., sex) or quantitative features (e.g., age) that individuals or populations might have or to defined sets or functions of these features (e.g., Medicare recipients). We call these definitions *pinpoint* concepts. Other concepts sort things into categories based on a set of criteria that are loose or hard to articulate precisely, where the members of the same category need not all share any defined set of features but rather have what Wittgenstein called "family resemblance." For reasons explained below, we call these family resemblance–like concepts *Ballung* concepts. "The number of people with diabetes mellitus," "mortality rate," and "the proportion of the population with impaired hearing" seem to be pinpoint concepts. Concepts with potential normative implications and those that have evolved from everyday concepts that serve a variety of purposes, such as "health," "disease," "human welfare," "human rights," and "poverty," tend to be Ballung concepts. An abundance of scholarship in the philosophy of medicine has attempted to define precisely a pinpoint concept of disease (Lemoine 2013), with several fairly successful yet distinct candidates surviving and no consensus in sight. One plausible explanation is that our everyday concept of disease is a Ballung concept that permits multiple rational reconstructions. Several diagnostic categories also seem to be Ballung concepts (e.g., "flu-like illness"), which denotes a family of infections resembling the flu.

The distinction here is not between natural features on the one hand and features that are socially constructed or dependent on social relations on the other hand. "Being a stepmother" depends on what social relations obtain but is a specific, unambiguous feature. "Race" is a paradigm for a concept generally taken to be socially constructed but that has been, in many societies, meticulously defined (e.g., based on region of origin) to allow labels to be applied unambiguously so that all the members of the same category do resemble each other in precisely the ways laid out in the definition (even though these labels may not have any other importance). Nor is the distinction we are trying to make one between the observable and unobservable. It would also be misleading to cast the distinction in terms of realism versus nominalism since family resemblance can certainly be real, and it might be a perfectly objective fact that individuals clump into categories according to these resemblances.

Otto Neurath[1] maintained that most concepts used in daily life are of the second type. He called them *Ballungen* ("congestions"),[2] as in the German "Ballungsgebiet," for a congested urban area with ill-defined edges. There is a lot packed into the concept. There is often no central core without which one does not merit the label, different clusterings of features among the congestion (*Ballung*) can matter for different uses, and whether a feature counts as being inside or outside the concept—and how far outside—is context and use dependent. We employ Neurath's word because other words more commonly in use throughout philosophy and the sciences, such as "umbrella concept" or "family-resemblance concept," have different meanings for different scholars in different fields.

Neurath's doctrines about Ballung concepts were influenced by Max Weber (1949). Weber argued that the study of society could probably not become a proper science because the hallmark of proper science is the use of precise, unambiguous concepts that figure in exact relations with one another. Physics, he believed, can pick and choose the concepts it studies to find such concepts, but the study of society has no such latitude since it is supposed to help us understand and manage the concepts we are concerned with in the conduct of life, few of which have the right character to participate in exact science. Concepts familiar to general society, such as "disability," "poverty," or "functional literacy," are bound to have multifaceted meanings, so offering a single, precise characterization is likely to sacrifice or alter aspects of their meaning. This may well be the case with respect to several medical concepts, such as "depression," "obesity," and "health."

It is essential that our measurement procedures measure the concepts we are aiming to measure, so the importance of definition cannot be overemphasized. Explicit definition is the most straightforward way to go and has become increasingly common in medicine, where explicit criteria for clinical categories are routinely decided at consensus conferences. When there is a well-articulated body of knowledge already accepted, implicit definition via the role the concept plays with respect to other concepts in a system of claims or axioms is the next tightest way to characterize a concept. This is the category that Northrop (1947) calls "concepts of postulation." Generally, we are not able to do this in the social sciences or medicine in part because we generally lack mathematical theories/models (though it is often not very easy in the natural sciences either). After all, part of the point of measuring a concept is to find out how it relates to other concepts. Usually, we need to start with some rough, defeasible characteristics of the concept and through a gradual back-and-forth process refine the characterization simultaneously while refining our procedures for measuring it and our claims about its relations to other concepts.[3] The fact that we often start with rough, open-ended characterizations does not imply that the concept in view is a Ballung concept since this is the historical trajectory of many natural science pinpoint concepts like "temperature" (Chang, 2004).

When our understanding of a concept and our knowledge of what other features might serve as good indicators of it are weak, we sometimes resort to one kind of explicit definition: operational definition. We point to a set of relatively well-articulated measurement procedures and define the concept in terms of them. The concept is then whatever it is that these procedures assign values to. The intelligence quotient (IQ) is the canonical example: "IQ," some maintain, "is just what IQ tests measure." Another example might be body mass index (BMI). We often speak of BMI as if it is a property of individuals ("the patient's BMI is 25"). But we cannot say much more about it than that it is the mass divided by the height squared.

Operationalization makes knowledge accumulation difficult. It becomes hard to justify that other procedures measure the same quantity since that requires defending the empirical hypothesis that the new procedures yield the same values as those that define the concept. We also often gain confidence in our measurement results and our characterization of the quantity by noting that different procedures for measuring it yield roughly equivalent results, which is difficult when quantities are defined operationally.

A good example of the difficulties of operational definition in medicine is provided by the *Diagnostic and Statistical Manual of Mental Disorders* (*DSM*), which is the premiere manual for diagnosing psychiatric disorders, particularly in North America. It operationalizes the diagnosis of mental disorders by providing lists of criteria (the presence of certain symptoms and behaviors, the absence of other disorders) that define the diagnosis in question. Operationalization supposedly increases interrater reliability of psychiatric diagnosis and aids in communication among health care professionals. But *DSM* categories are less helpful in research settings presumably because there is much causal heterogeneity underlying *DSM* disorders. Part of the reason for this heterogeneity is that symptoms and behaviors are variably caused and multiply realized. Another explanation for the underlying heterogeneity is that many *DSM* diagnoses are polythetic (i.e., a patient must satisfy a certain number of criteria, none of which are individually necessary). The way that mental disorders are operationalized by the *DSM* creates challenges for research and the accumulation of knowledge. It becomes difficult to create scientific models of mental disorders and to design treatments that have large effect sizes in *DSM*-defined categories of patients.

No matter which of the two types[4] of concept—pinpoint or Ballung—we consider, in characterizing the concept, we are usually pulled in two different directions: generality and fitness for purpose. Making concepts more precise can make them more fit for purpose, but it proliferates concepts and measures. Moreover, as with concepts defined operationally, purpose-built concepts make the accumulation of knowledge difficult, so we are often reasonably pulled to rely on more general concepts that have poorer fit and hope that results established in one situation are relevant for other situations.

Representation

Systems of Representation

Representation of pinpoint concepts is usually done using a metrical system with an underlying mathematical structure. Stevens (1951) enumerated four kinds of representations. The representation may simply be a record of the number of tokens of a concept (e.g., proportion of a population that is male or has a particular disease [often called *nominal* measures[5]]); the representation may order tokens (e.g., rank hospitals on their reputation [*ordinal* measures]); it may order tokens on a scale with equal intervals (e.g., blood glucose [*interval* measures]); or it may order tokens on a scale with equal ratios and a true zero point (e.g., perceptual scales of loudness [*ratio* measures]).

Although the distinction between pinpoint and Ballung concepts is not sharp, it is a useful distinction to keep in mind in thinking about representation. For instance, an interval or ratio scale would be inappropriate for a Ballung concept. For Ballung concepts, there are three common strategies for representation. One strategy is to shed much of the original meaning and zero in on more precisely definable features from the congestion that constitutes the concept. A second strategy is to represent the concept with a table or vector of features laying out the dimensions along which the family resemblances in question lie. A third strategy aims to compromise between the advantages and disadvantages of these two previous strategies by starting with a set of different indicators but then amalgamating them into a single index number.

An example of the first strategy is provided by Sophia Efstathiou's (2012) discussion of the controversial Ballung concept "race" and its introduction into different medical contexts. Efstathiou points out that the kind of variation that "race" picks out for epidemiologists and the kind it picks out for geneticists is different in type. Epidemiologists mostly care for variation in common health outcomes and risk factors for diseases; "race" in epidemiologists' speak is defined in terms of regional heritages, which may then be associated with disease risks. Geneticists care about markers in the genomes of individuals, and they look for interesting patterns at the level of molecular variation; their characterization of "race" maps onto sets of genetic polymorphisms. This is a case where a loaded Ballung concept, "race," is used in the service of different scientific research projects. Most of its content is discarded in each context, and a narrower and more precise characterization is fitted to the scientific questions being asked. This leads to a proliferation of concepts, each fit for a different discipline and purpose. Different concepts may share the same name, but they are not equivalent; thus, results established in one setting cannot be transferred to the others.

Medical "quality of life" (QOL) is a Ballung concept par excellence. It is made up of many domains, each with its own characterization, and there is no specific delineation of its borders. The concept has thus spawned a proliferation of measures. While each individual domain, such as depression, anxiety, pain, or mobility, may be well measured in principle (that is, its characterization, representation, and procedures are well worked out), differing domains may be included in the concept depending on the purpose to which the QOL measure is put. A treatment for a certain neurological disease might aim to improve mobility or cognition, and a treatment for pain disorder will aim to alleviate pain. The goal of improved QOL may be common to the different studies, but different measures may be used for QOL depending on the aims of the treatments (e.g., improved mobility vs. reduced pain). Consequently, it would be impossible to compare QOLs across different kinds of treatments unless it were clear that the domains underlying the QOL indices were the same.

The second strategy is to represent Ballung concepts with a table or vector of indicators, illustrated by ratings of hospital quality. A well-known measure of hospital quality (Hill and Winfrey 1996) has three tiers: *structure*, *process*, *and outcomes*. *Structure* is represented by such things as "ratio of interns and residents to beds," "ratio of registered nurses to beds," "ratio of board-certified doctors to beds," and an index of available technology. For some specialties, there is an added measure of number of procedures performed in a year (volume) and for others, the availability of special services, such as discharge or planning services. *Process*, which is difficult to measure directly, is measured by reputation, as determined by a survey of specialized physicians as a proxy to quality of care. *Outcomes* are measured by such things as risk-adjusted mortality rates, readmissions, and infection rates. The actual measures used have evolved since the beginning of the ratings in 1993, but they retain their essential tiered structure.

This strategy comes with drawbacks. Tables and vectors do not make for easy comparison either across time or across groups. For example, *Healthy People, 2020* (www.healthypeople.gov/) has twenty-six Leading Health Indicators for population health, increased from ten indicators in *Healthy People, 2010*, with progress to be tracked over a decade. Comparison across populations is not impossible, though. High rankings on all indicators orders a group above one with low rankings on all indicators, and there may be further reasonable ways to rank groups where differences on most indicators are large or enough indicators differ in the same direction. Nevertheless, at best we can expect only a partial ordering. That is not a problem with the measure. It is often, as Weber urged, the problem with the concept in which we are interested. There simply is no fact of the matter about which country among a large number of European countries with mixed results has a healthier population.

Sometimes we try to "solve" the problem by collapsing the indicators into a single index, generally by defining some weighting scheme to produce a single outcome. The Apgar score is an example; it is an immediate postnatal screening tool to help determine how well a newborn tolerated delivery and how well she or he is faring outside the womb. The Apgar score ranges from 0 to 10 and is calculated by adding subscores from five equally rated assessments: breathing effort, heart rate, muscle tone, reflexes, and skin color. As a further example, the Montreal Cognitive Assessment (MoCA) tool screens for mild cognitive impairment by testing cognitive functioning in several domains: visuospatial/executive, naming, memory/delayed recall, attention, language, abstraction, and orientation. Again, subscores in each domain are added to generate an overall score, but the domains in MoCA are not equally weighted.

Aggregating indicators into a single index has obvious advantages and disadvantages. On the one hand, it makes comparisons and accumulation of knowledge easier. On the other, the choice of weighting scheme is often underdetermined and sometimes downright arbitrary, which opens the possibility of cherry picking just the right weightings to get desired results (e.g., in a clinical research study with lucrative implications). In egregious cases, some of the indicators included in the composite measure might not be relevant to the outcome we care about. For instance, Goldacre (2012) describes the influential UKPDS trial that analyzed the effect of blood sugar control on diabetic endpoints. The trial showed a 12 percent reduction in the composite endpoint due to intensive blood sugar management in patients with diabetes. The composite endpoint included important outcomes, such as sudden death, heart attack, and stroke, and indirectly relevant outcomes, such as renal biomarkers. On closer analysis, most of the 12 percent improvement in the composite outcome was due to a reduction in the number of patients referred for laser treatment for damage to the microvasculature of the retina rather than due to improvement in the most important cardiovascular outcomes (meanwhile, there was no significant change in the number of patients experiencing vision loss).

There are further problems beyond issues of how the weights are selected. When the original concept is a Ballung concept, this strategy amounts to constructing a new, more manageable concept rather than informing us about the original. Of what interest is this new concept? What purposes are served by measuring it? It may be that the new concept is useful for scientific theorizing, for prediction, or for explanation—for enterprises that rely on principles involving concepts that are precise and unambiguous. In this case, it is probably most useful to treat it no longer as a Ballung concept but rather as a pinpoint concept since playing a role in a network of predictive principles is one of the chief grounds on which we judge concepts to pick out specific, precise features.

Conversely, it can be misleading to represent pinpoint concepts by sets of indicators or indices. Sometimes we can find no procedures that will tell us about a pinpoint concept directly, so we must resort to measuring the concept in a host of indirect ways, none of which suffices to zero in on it sufficiently reliably or precisely. In this case, good practice would be to report the array of results. However, to represent such a concept in a theoretical structure in indirect ways risks losing the opportunity of laying out any exact relations in which it figures. Doing so blurs the line between what is vague in the world and what we are uncertain of, and blurs it in an unhelpful way.

Representation Theorems

In this chapter we offer a general account—a theory of the nature of measurement, especially in the social sciences and medicine.[6] Sometimes the term *theory of measurement* is used more narrowly to refer to concerns about the connection between the characterization and representation; and sometimes even more narrowly, to the abstract characteristics of the formal method of representation.[7] For proper measurement, these abstract characteristics of the formal representation of a concept must reflect and be warranted by the characterization of what the feature or category to be represented is. For instance, if a feature is to be represented on a scale of 1 to 10, that scale should be treated as a pure ordering (an ordinal as opposed to a ratio measure) if nine units does not equal three times the amount of the quantity possessed by tokens with three units. Thus, the central task in designing a good measure according to the narrow sense of measurement theory is to provide a "narrow" *representation theorem* to show that the representation proposed has formal, abstract features appropriate to the concept as it has been characterized.

We endorse the demand for representation theorems of this sort. However, we want to underline the need to produce arguments that address the more substantive aspects of the representation. For instance, suppose our procedures dictate the use of a mercury thermometer to measure temperature. This implies that temperature is represented by the height of a column of mercury, which in turn is formally represented on an interval scale. To justify the representation, we must show that an interval scale is appropriate to the kind of thing our characterization says temperature is. But that is not enough. We must also show why readings of the height of a column of mercury can indicate temperatures in the way proposed. That will involve many substantive assumptions—such as the assumption that mercury expands uniformly with temperature.[8] The representation theorem makes these assumptions explicit and lays out the argument that shows that column height and temperature are indeed related as presupposed in the procedures. The measure itself can be no more warranted than the assumptions required for the proof.

Unfortunately, representation theorems are often lacking in the case of clinical research measures. Stegenga (2015) discusses the example of the Hamilton Depression (HAMD) rating scale. The HAMD questionnaire rates the severity of depression on a scale from 0 to 52 through scoring patient responses to seventeen questions. One question probes the degree of suicidality, with responses rated from 0–4 (0 = suicidality absent, 4 = attempts at suicide). In contrast, there are a total of six points available to quantify the degree of insomnia and four points available to quantify the amount of fidgeting. Thus, a patient who had attempted suicide might score the same with respect to these three elements of depression as a nonsuicidal, somewhat fidgety patient with mild insomnia. As Stegenga observes, an antidepressant might even improve HAMD scores in a clinical trial by causing patients to sleep better and fidget less (a generic sedative could also achieve this outcome). It seems unlikely that any sound representation theorem supports the use of the HAMD scale as a representation for the concept "severity of depression."

One place where the need for representation theorems looms large is in the construction of index numbers by weighting different indicators. There is, as we said, a great deal of pressure to do this since a total ordering of the tokens measured will then be possible, whereas with tables or vectors, usually at best only a partial ordering is possible. But in this case, there should be good arguments that the weightings are appropriate to the concept to be measured and that the final representation does not imply features that the concept does not have. Often, weightings are not explicitly mentioned in medical measurement, which is tantamount to tacitly applying equal weightings to indicators. In quantifying the amount of morbidity for a patient, for example, we may simply count chronic diagnoses. This scheme uses a ratio scale that counts a patient with two diagnoses as having twice the amount of morbidity as a patient with one diagnosis. Each diagnosis tacitly receives an equal weighting even though some diseases can have a greater impact on mortality, quality of life, and other outcomes compared with other diseases.

Measurement Procedures

We most commonly think of measurement in science in terms of the procedures we carry out to assign measurement values to tokens in the world. In setting up these procedures, effort should be made to ensure that they are both *accurate* and *precise*. In common parlance, "precision" is often conflated with "accuracy." Here is one way to regiment the use of the words: *accuracy* is about whether measurement results agree with the true values or locate individuals in the correct category;[9] *precision* indicates how specific a measurement result is.

Where a genuine quantity is measured, the observations are often done with an instrument that is calibrated to the metrical system that represents the quantity, such as a ruler or thermometer or by a simple counter. The observations can be transformed into various metrical systems by algorithms, such as converting feet into meters or degrees Fahrenheit into degrees Celsius. In many cases, as we noted, these instruments do not look at the quantity directly but rather rely on a preestablished connection between the quantity to be measured and another more directly observable quantity as, for instance, with the mercury thermometer. Similarly, we use pulse to measure heartrate, assuming that the number of arterial pulses is equal to the number of ventricular contractions, and blood pressure using a blood pressure cuff. Sometimes the more immediately observed quantity will be a cause of the targeted concept, sometimes an effect, and sometimes the two are correlated for some other reason. What matters is that the two quantities be linked by reliable regularities. Laying out and defending these regularities is one of the central tasks in designing a measurement procedure.

This gives rise to what is sometimes called "the problem of nomic measurement" (Chang 2004, 59). To be confident that the mercury thermometer measures temperature accurately, we must be confident that mercury expands uniformly with temperature. But to establish this empirical regularity we need an independent and accurate method of measuring temperature. This problem of justification is common to all measurement methods based on empirical laws. There are several obvious ways to circumvent this problem. First, we can determine the values of the quantity we want to measure, such as temperature, by another method (this only postpones the problem since now that other method needs to be justified). Second, we can derive the empirical law from a general theory. This is not straightforward either since the theory relied on must be empirically justified, which can be especially difficult in the social sciences and medicine, where few theories are accepted uncontroversially.

Both strategies are routinely employed in evaluating the accuracy of diagnostic tests in medicine. We often measure the accuracy of a test by comparing its performance to a gold standard test that is assumed to be nearly perfectly accurate. The accuracy of a d-dimer blood test for detecting a blood clot in the lungs can be measured by comparing its performance against a CT angiogram. The gold standard diagnostic test is often chosen based on theory.

CT angiogram is considered the gold standard for detecting blood clots in the lungs because it visualizes clots radiographically with great resolution, an assumption that depends on both medical science and physical theory.

A mild version of operationalism can also be an attempt to circumvent the regress. If empirical concepts are defined by well-specified measurement operations, observational data can be fixed without reference to theories and be made secure even while theoretical concepts and laws fluctuate and develop. This tactic brings with it all the problems we have discussed of narrowness, comparability of results from different methods, and the danger that we are no longer talking about the concept that we started out to study.

Another tactic is to look for operations that depend on relatively uncontentious empirical principles or that should give near enough the same results across the range of competing empirical principles that are deemed plausible. Whether these operations are available depends on the circumstance. Herbert Feigl (1970) argued that our most basic measurement operations are grounded in middle-level regularities that seem to have a remarkable degree of stability, such as Archimedes's law of the lever and Snell's law of refraction. Again, finding these regularities seems especially problematic in the social sciences and medicine, although in physiological measurement they often exist (e.g., measuring heartrate using arterial pulse based on the typically regular association between ventricular contraction and arterial pulsation).

For psychological concepts that are messy and in principle unobservable, Campbell and Fiske (1959) advocate a multitrait, multimethod approach to validation. Concepts can be accepted only when they can be measured by several different methods and with different representations. This is an example of the back-and-forth process of refinement—often called "triangulation"—among characterization, representation, and design of procedures. Much work needs to be done before a proper measure, where all three components fit appropriately, is arrived at, and much substantive knowledge can be involved in the process.

Coherence along a variety of desiderata seems to provide the best solution in practice to the problem of nomic measurement in medicine as well as in the natural and social sciences. Does the quantity as measured by the proposed method behave as it is expected to? Do the results cohere with those of other reasonably defended methods? Do the empirical principles needed to support the method cohere with other reasonably justified empirical principles and theories?

Another common strategy is to provide a vector or table of results from different methods, with perhaps some attempt to describe the spread of results statistically. It is important to keep in mind, however, that in this case there is a different reason for using vectors or tables than with Ballung concepts. The difference in reason can have important consequences for how we use the information thus presented and for how we proceed to develop our science and measurement procedures since in the first instance, we suppose that there is a single correct value to be ascertained and in the latter case of Ballung concepts, we do not.

Just as the representation of a concept must match the concept, the procedures that assign values must match its representation. Mere counting of number of obese individuals in a population may be adequate if we represent "obesity" as a dichotomous category, but it is a poor procedure for assigning a value to the amount of obesity for a measure that is sensitive to degree (e.g., BMI). Measurement systems for subjective phenomena, that is, those for which there are in principle no appeals to consensus of external observation, are particularly difficult because there is no direct way of knowing that the subjective judgments are using the same scale.[10] For subjective phenomena, measurement typically starts with observations that depend on responses from individuals to (more or less) common stimuli. The responses are then represented in some metrical system with (more or less) well-defined properties. Sometimes these observations have a one-to-one relation between the response and the metric, as in the case of the "just noticeable difference" (JND) measures of sensation, and others, by response categories labeled with vague quantifiers, such as "not very often," "often," or "frequently," which are then mapped onto numeric values with only ordinal properties. Often, however, the observations are combined in some (more or less) well-specified way and put forth as a measure of a complex subjective phenomenon, such as an attitude or an illness experience. For an example of the development of a measure of psychological well-being, see Bradburn (1969).

The procedures used to measure a social or medical concept may end up producing a measure that does not correspond to the way the concept is meant to be understood. An arguable example is the measurement of mildly elevated blood pressure, moderately low bone density, or moderately high blood cholesterol as disease ("hypertension," "osteoporosis," and "hypercholesterolemia," respectively).

The borders of disease categories often change, leading to changes in rates of disease. For example, while the American Heart Association (AHA) has long considered systolic blood pressure less than 140mmHg and diastolic blood pressure less than 90mmHg as normal for adults, in 2003 the National Heart, Lung, and Blood Institute (NHLBI) set new clinical guidelines lowering the standard normal readings to a systolic pressure equal to or less than 120 and a diastolic pressure equal to or less than 80. Thus, the official statistics using AHA cutoffs report higher rates of normotension (normal blood pressure) than would be found with the NHLBI guidelines. The extent to which physicians have used the new guidelines to start treatment for patients viewed as having normotension according to the AHA guidelines is unknown. The public who are attentive to their health may be confused about whether they have normal blood pressure or not.

Even when there are good measurement properties, different operations may be used for different purposes. Consider again measures of QOL used to evaluate the effectiveness of different treatments. One set of procedures with good measurement properties is the Patient-Reported Outcomes Information System (PROMIS) developed by the National Institutes of Health (NIH). This effort arose out of concern for the large number of items used by various researchers to measure aspects of medical QOL. Many of these measures had poor measurement properties: it was unclear what characterization they were meant to be procedures for, what their relations were to other proposed procedures for measuring the "same" or "similar" concepts, or how they correlated with other concepts of interest. As part of a larger NIH effort to improve measurement, the PROMIS project, through elaborate review processes, has categorized concepts of interest, revised items, and submitted them to scaling procedures to produce measures that have desirable properties. The scales have been standardized on large, general populations and some specialized clinical populations. That the scales are accurate representations of the concepts was established by a large and complex series of clinical studies in which the scales were correlated with clinical and patient-reported assessments. These procedures are used to establish that the scale values map onto clinically meaningful assessments in regular ways. The measures then become the criteria for evaluating changes in treatment outcomes. But as noted above, different measures may be used to characterize QOL outcomes depending on the purpose of the interventions, thus making it difficult to compare levels of QOL among different groups or over time.

CONCLUSION

A good measure satisfies three requirements: (1) We have a characterization of the concept that identifies its boundaries and fixes what tokens belong to the concept and what tokens do not; (2) we have a metrical system that appropriately represents the concept as characterized; and (3) we have rules for applying the metrical system to individual tokens to produce measurement results. Only when the characterization, representation, and procedures are well specified and shown to mesh properly has a good measure been achieved.

We distinguished between pinpoint concepts that refer to a single quantity or category that can be precisely defined, and Ballung concepts, which refer to things that are loosely related but for which the boundaries of the concept are not clear. These kinds of concepts need to be treated differently not only with respect to characterization but also when it comes to representation and the design of procedures.

The use of concepts for different purposes often leads to changes in definition, representation, and/or procedures that disrupt the possibility of comparison and knowledge accumulation but that often make the measures more appropriate to the aim they are supposed to serve.[11]

NOTES

1. Neurath was a socialist, sociologist, philosopher, one of the founding members of the Vienna Circle, and spearhead of the unity of science movement of the 1930s.
2. See Cartwright et al. (1996) for discussion and references.
3. For an example of this back-and-forth process from the natural sciences, see Chang (2004).
4. There is a third type of concept important in the sciences that we will not discuss, what might be called "concepts of pure understanding." They are useful for understanding, for representation (often mathematical representation), or for organization but not for literal description.
5. Not to be confused with taking a *nominalist* metaphysical stance towards them.
6. For a related general account see Chang and Cartwright (2008).
7. See Suppes (1998) for an accessible introduction.
8. For a fascinating study of the long effort to get characterization, representation, and procedures to mesh well in the case of temperature, we again recommend Chang (2004).
9. For nominalists, accuracy can be taken to indicate that the term has been applied in accord with all the accepted norms.
10. Or at least some transformations of the same scale.
11. Nancy Cartwright's work on this chapter was sponsored by grants from the Templeton Foundation, the British Academy, the UK AHRC, the UK LSE ESRC Centre for Climate Change, Economics and Policy, and the Durham project *Knowledge for Use* (*K4U*), which has received funding from the European Research Council (ERC) under the European Union's Horizon 2020 research and innovation program (grant agreement No 667526 K4U). The content reflects only the author's view, and the ERC is not responsible for any use that may be made of the information it contains.

REFERENCES

Bradburn, Norman. 1969. *The Structure of Psychological Well-Being*. Chicago: Aldine Publishing Co.

Campbell, Donald T., and Donald W. Fiske. 1959. "Convergent and Discriminant Validation by the Multitrait-Multimethod Matrix." *Psychological Bulletin* 56: 81–105.

Cartwright, Nancy, Jordi Cat, Lola Fleck, and Thomas E. Uebel. 1996. *Otto Neurath: Philosophy between Science and Politics*. New York: Cambridge University Press.

Chang, Hasok. 2004. *Inventing Temperature: Measurement and Scientific Progress*. New York: Oxford University Press.

Chang, Hasok, and Nancy Cartwright. 2008. "Measurement." In *The Routledge Companion to Philosophy of Science*, edited by Stathis Psillos and Martin Curd, 367–75. London and New York: Routledge.

Efstathiou, Sophia. 2012. "How Ordinary Race Concepts Get to Be Usable in Biomedical Science: An Account of Founded Race Concepts." *Philosophy of Science* 79: 701–13.

Feigl, Herbert. 1970. "The 'Orthodox' View of Theories: Remarks in Defense as Well as Critique." In *Analyses of Theories and Methods of Physics and Psychology*, edited by Michael Radner and Stephen Winokur, 3–16. Minneapolis: University of Minnesota Press.

Goldacre, Ben. 2012. *Bad Pharma: How Drug Companies Mislead Doctors and Harm Patients*. New York: Faber and Faber.

Hill, Craig, and Krishna Winfrey. 1996. *The 1996 Index of Hospital Quality*. Chicago: NORC at the University of Chicago.

Lemoine, Maël. 2013. "Defining Disease Beyond Conceptual Analysis: An Analysis of Conceptual Analysis in Philosophy of Medicine." *Theoretical Medicine and Bioethics* 34: 309–25.

Northrop, F. S. C. 1947. *The Logic of the Sciences and the Humanities*. New York: Meridian Books, Inc.

Stegenga, Jacob. 2015. "Measuring Effectiveness." *Studies in History and Philosophy of Biological and Biomedical Sciences* 54: 62–71.

Stevens, Stanley S. 1951. "Mathematics, Measurement, and Psychophysics." In *Handbook of Experimental Psychology*, edited by S. S. Stevens, 1–49. New York: Wiley.

Suppes, Patrick. 1998. "Theory of Measurement." In *The Routledge Encyclopedia of Philosophy*, edited by Edward Craig. London: Taylor and Francis. https://www.rep.routledge.com/articles/thematic/measurement-theory-of/v-1.

Weber, Max. 1949. "'Objectivity' in Social Science and Social Policy." In *The Methodology of Social Sciences*, translated and edited by E. A. Shils and H. A. Finch, 50–112. Glencoe, IL: Free Press.

Chapter 6

Psychological Measures, Risk, and Values

Leah M. McClimans

Person-centered measuring instruments, the kind currently in vogue for use in medical and public health research, are a form of psychological measurement. To understand better their possibilities and limitations it is useful to turn to the literature in psychometrics, the study of the theory and methods of psychological measurement. Leading trade journals for psychological measurement, such as *Psychometrika*, *Theory and Psychology*, and *Measurement*, publish philosophical debate, debates conducted mainly by psychometricians. One aspect of this debate centers on the ontology of psychological attributes and what measurement theories and methods befit them. Interestingly, most of the psychometricians participating in these debates agree in general that only a realist ontology is suitable for psychological measurement (Borsboom 2006; Maul 2013; Michell 2005). Their debate, in terms of ontology, is thus less directed toward one another than it is toward psychologists and others who continue to employ measuring instruments built out of theories and methods that cannot sustain a realist ontology (Borsboom 2006; Michell 1999).

In this chapter I discuss two perspectives on what is at stake in the choice between measuring instruments. Psychometricians who argue for realism tend to rely on epistemic values, such as rigor and truth, to make their case. On this view, the integrity of psychological measurement as a science is what is taken to be at stake in the choice of measure. Contrasting this viewpoint, I argue that nonepistemic ethical values, such as trust and harm, are equally important in measurement choice and particularly figure into the person-centered psychological measures often used in health care. I further argue that because these measures require nonepistemic considerations in their development and application, they challenge a strict insistence on realism in measurement. What is at stake is not the integrity of science so much as the usefulness of these measures as tools for medical and public health research.

ONTOLOGY AND PSYCHOLOGICAL MEASUREMENT

There is much debate in philosophical discussions of psychological measures (e.g., what kind of measurement methods should be used to produce psychological measures, the role of validity and whether psychology is even capable of genuine measurement). But standing alongside these debates is an emerging consensus among psychometricians that psychological measures require realism. Different proponents of this view make different arguments for it, but to provide a flavor of this position, consider two leading proponents, Joel Michell and Denny Borsboom.

In the same year that Michell (2005) published his article "The Logic of Measurement: A Realist Overview," Borsboom (2005) published his monograph *Measuring the Mind.* Both publications argue for realism in measurement and situate this position as an alternative to antirealism (e.g., mathematical theories of measurement, such as representational measurement theory [RMT], and operationalist theories of measurement, such as classical test theory [CTT]). Michell advocates for entity realism, which has at its core a correspondence theory of truth whereby a proposition is true if and only if circumstances are as the proposition states (Michell 2005). Thus, if the scale says I weigh 59 kg, then this is true only if I exist at a particular time and space, the attribute weight exists, and my weight is 59 kg. To say that weight exists is to say that it exists at a particular time and space as a feature of something else (e.g., a person, an object). Moreover, attributes such as weight exist as a range, for instance, I might weigh 59 kg now and 61 kg in a year.

Michell's realist ontology strives to align measurement with scientific discovery; indeed, he argues that only realism can do justice to scientific discovery (Michell 1997). For Michell, measurement fundamentally relies on the discovery of ontologically real quantities. Accordingly, metrologists, like other scientists, make hypotheses, gather evidence, and then use that evidence to draw conclusions about the likelihood of their hypotheses. In the case of measurement, the basic hypothesis should always be the same: attribute X is a quantity. This hypothesis refers to the internal structure of an attribute. For an attribute to be quantitative, it must be structured such that the values of the variable stand in certain algebraic relations to one another. Specifically, the relations must be ordered (e.g., transitive and additive), and conforming to the properties of addition (e.g., commutativity). Another way to put this point is that different instances of the same attribute must sustain ratio relations (Michell 2005). On Michell's realist perspective, these ratios are real numbers that represent different levels of the attribute. The job of metrologists is to gather evidence (e.g., concatenation or conjoint additivity) to determine whether this hypothesis is true and thus whether measurement of a specific attribute is possible.

Like Michell, Borsboom's realism concerns both the independent existence of attributes and a truth correspondence between the affairs postulated by a theory and those in reality. But Michell and Borsboom differ in that the importance of realism for Borsboom is not the discovery of a quantity. For Borsboom measureable attributes can be either quantities or qualities; what matters for measurement is whether an attribute is causally relevant. Specifically, he is concerned that there is a causal connection between an attribute represented by a latent variable and between-subject responses to items.[1] To this end, his argument for realism invokes latent variable theory.

Latent variable theory, or modern test theory, as it is sometimes called, is a measurement theory that uses mathematical models to hypothesize the attribute and parameters needed to explain the empirical data from psychological tests. As we saw above with Michell, measurement is here again taken to be an instance of scientific discovery. We use these models to test their adequacy against observed test data (i.e., the extent to which the observed data "fit" the predictions of those responses from the latent variable model, within acceptable uncertainty). If the model fits, then we have some evidence to suggest the observed data are a function of the model. But model fit is underdetermined (i.e., multiple models may fit the observed data within acceptable uncertainty [Borsboom 2005]). Thus, "fit," although necessary, is not sufficient to pick out the correct latent trait for any data set.

Borsboom's argument for realism is that some ontology is needed in addition to "fit" to motivate the choice of model in latent variable theory. In practice psychologists using latent variable theory choose reflective models to explain the empirical data from tests. Borsboom argues that reflective models presuppose realism. Reflective models specify that the pattern of covariation between observed item responses (e.g., answers to reading questions) can be explained by a regression on the latent variable (e.g., reading ability). The idea is that item responses vary as a function of the latent variable (i.e., differences in reading ability affect differences in item responses). Borsboom contrasts reflective models with formative models (Borsboom 2005). With formative models, which are common in sociological and economic modeling, the latent variable is regressed on the observed item responses. Put differently, responses to questions about, for instance, income and education affect the latent variable (e.g., socioeconomic status). In reflective models the latent variable (e.g., reading ability) is understood as determining our measurements, and in the formative model, the latent variable (e.g., socioeconomic status) is a summary of them (Borsboom 2005).

In principle Borsboom suggests there is no reason why psychologists should choose one model over the other; nonetheless they consistently choose reflective models. Why? His answer is that in this choice, psychologists reveal an ontological commitment to entity realism; the choice of a reflective model presupposes an agent-independent latent variable that causally affects observed responses between subjects (i.e., at the population level). For the purposes of this chapter, let us accept the implications Borsboom draws from the choice of a reflective.

Michell and Borsboom differ in their accounts of measurement; yet they share a commitment to entity and theoretical realism as well as a commitment to scientific inquiry as orientated toward discovery. For Michell, measurement is the estimation of the ratio between two instances of a quantitative attribute, and because quantitative attributes must first be discovered before measurement can take place, this requires that the attributes are ontologically real. For Borsboom, measurement also requires an ontologically real attribute with a determinate structure, but that structure may be qualitative or quantitative. Additionally, Borsboom requires a measuring instrument that is sensitive to variations in the attribute and able to reflect those variations.

No matter what we think about the success of these arguments for realism and the accounts of measurement they underpin, I want to draw attention to the ontological position with which these arguments are meant to contrast: antirealism. Antirealist positions are taken in this debate to be represented by two distinct measurement theories: (1) representational measurement theory (RMT) and (2) classical test theory (CTT). Although RMT is more widely known within the philosophical literature, my focus is on CTT because this is the measurement theory most widely used in psychology. Nonetheless, let me briefly touch on RMT first.

Representational Measurement Theory

RMT seeks to map numerical relations onto qualitative empirical relations in such a way that the information in the empirical set is preserved in the numerical set. The creation of this homomorphism is a measurement scale. Consider for instance the empirical set of rigid rods and the relations between them (e.g., longer than). Representational measurement's goal is to specify numbers and mathematical relations (e.g., addition) that map onto the empirical structure of the rigid rods. Much of the literature of RMT concerns the identification of axioms that hold between objects, the development of representational theorems that specify when a homomorphism is possible, and uniqueness theorems that identify what kind of measurement scale a specific measurement procedure will produce.

RMT is concerned with determining the relations that must manifest in the empirical data to construct a measurement scale. As such, RMT is often criticized for its overly rational orientation and inability to engage with everyday issues in applied measurement (e.g., measurement error and uncertainty, calibration, reliability, and so on) (Borsboom 2005; Heilmann 2015; Tal 2013).

In the same vein, it is criticized for being overly abstract and thus not engaged with concrete scientific inquiry. As Borsboom (2005) notes, RMT does not hypothesize theoretical constructs or latent variables; the measurement scales created using RMT require only basic empirical relations, logic, and mathematics. It implies an antirealist ontology closely related to logical positivism by linking empirical (observed) relations to numerical (theoretical) relations via axioms and theorems, which for Borsboom resemble correspondence rules (Borsboom 2006; Michell 2005).

Classical Test Theory

Although RMT does not serve to inspire many contemporary psychological measuring instruments, CTT does. In fact, CTT is the dominant measurement paradigm within much of psychological measurement including clinical outcome assessments (COAs) (Borsboom 2006; Cano and Hobart 2011). CTT turns on a simple model where an observed score (O) (i.e., the empirical data acquired after respondents complete a test or questionnaire) is equal to a person's true score (T) plus uncertainty, commonly termed random error (E); thus $O = T+E$.

When using CTT the value of the true score is taken to be a theoretically unknown value, which is assumed to be constant, and the observed score is assumed to be a random variable, which produces a bell-shaped curve around the true score. The error score is taken to have an expectation value of zero. The idea here is that as the number of observations (i.e., administrations of the test or questionnaire) increases, the random errors will tend to cancel one another out; thus, the mean of the observations is taken as an estimate of the true score. To acquire an empirical value for T in the context of COAs, a person must be measured repeatedly on a scale (fill out the items of a questionnaire), and each observation (individual items or repeated administration of the same questionnaire) must be independent of the others (Hobart and Cano 2009).

The problem with these requirements is that in much of the behavioral sciences, they are not met. For instance, repeated administrations of a COA are not independent of one another. Respondents remember the questions from previous administrations and reevaluate their answers considering them. Moreover, COAs often do not function as a successive series; rather they function as measurements taken at a single point in time (e.g., to determine one's physical functioning three months post-operation). Third, the interpretation of the observed score as an estimate of the true score significantly depends on the assumption of a continuous variable (e.g., distance) with a normal probability distribution. But many of the variables in the context of COAs (e.g., physical functioning) are categorical, not continuous, as the responses elicited from respondents to individual questions take a limited number of values (e.g., strongly agree, agree, disagree, strongly disagree).

These difficulties, as well as others, are well known (Borsboom 2006; Cano and Hobart 2011). Typically, a thought experiment is given to manage the first two: imagine the person being measured is brainwashed in between a series of administrations (Lord and Novick 2008). This thought experiment renders, by definition, a series of administrations, and the brainwashing renders those administrations independent of one another. The third difficulty is often dealt with by simply ignoring the categorical nature of the data elicited from individual questions and assuming that the variable approximates continuity given a large enough number of possible values that can be derived from combinations of responses to different questions.

One drawback of this thought experiment is that it renders CTT unfalsifiable (Borsboom 2005; Hobart et al. 2007). Borsboom (2005) goes so far as to call it a "tautology." This is because CTT is rooted in the theory of errors. Within this theory, the idea that random errors will cancel one another out in the long run (i.e., the error score will have an expectation value of zero) is an empirical assumption (Borsboom 2005). That an observed score estimates the true score is a hypothesis contingent on this empirical assumption. But in the context of psychological measurement, there are no empirical grounds for making these assumptions since (1) measurements are not taken in a series and (2) even if they were, they are not independent of one another. To put CTT into practice, the error score must be assumed to have a zero expectation value, and as a result the true score is simply defined in terms of the mean observed score. In practice, the model reduces to O = T.

The claim that CTT is a form of antirealism derives from this simplified model, and at least for Borsboom (2006), it evokes an antirealist version of operationalism.[2] Operationalism is best known through the work of Percy Bridgman (1927, 5), who wrote, "In general, we mean by any concept nothing more than a set of operations; the concept is synonymous with the corresponding set of operations." CTT is open to an operationalist interpretation since the observed score, which is the result of a particular instrument's "operations" essentially defines the meaning of the attribute in question through its equivalence with the true score. Indeed, we can see operationalism at play in the context of COAs where we find a proliferation of measures targeting quality of life or subjective health status.

Epistemic Risk and Psychological Measurement

CTT is widely criticized as a theory for psychological measurement, but in this section I want to focus specifically on the kind of criticisms that those who advocate for realism in measurement make to illustrate the kind of risk they understand CTT to proliferate and thus the values they take it to represent.

In his article "Attack of the Psychometricians" Borsboom (2006) analyzes (and laments) various factors that have stalled the integration of psychometrics with psychology. One might imagine that because psychological measurement is a large part of psychology, psychometrics should be an integral part of it. Yet the advances in modeling that psychometrics has achieved in the last century have not been taken up by psychologists who develop and use psychological measures. In this paper, Borsboom is interested in why this integration has not occurred. He cites the prevalence of CTT in psychology textbooks, operationalism as a popular theory in psychology, and a lack of interest and training in the math necessary to understand and use the developments in psychometric modeling. Of interest to this chapter is, why does Borsboom think these advances in psychometrics *should* be integrated into psychology? He provides two different explanations. First, an integration is necessary for the progress of psychology as a science. Second, current instruments are not fit for purpose, and because they directly affect individuals' lives, we have a social obligation to improve them.

Jeremy Hobart and colleagues (Hobart et al. 2007) elaborate on these explanations in their paper "Rating Scales as Outcomes Measures for Clinical Trials in Neurology: Problems, Solutions and Recommendations." They argue that using new psychometric methods (i.e., latent trait theories) improves measuring instruments' scientific rigor and thus the chances of coming to a correct conclusion about the effect of a disease and the efficacy of a treatment given clinical change. They examine two methodological limitations of CTT to make these points.

Ordinal vs. Interval Measurement Scales

CTT yields measurement scales at the ordinal level. It can be difficult to interpret the significance of (e.g., clinical change) from ordinal-level measurement scores. For instance, if a group scores 15 on the Beck Depression Inventory and then scores 19, their depression has gotten worse, but how much worse? Does this change indicate a need for clinical intervention? Or is the change nominal? If another group changes from a score of 41 to 45, their depression has also gotten worse, but is this increase comparable to the first group? Because ordinal scales lack a consistent unit, these questions are difficult to answer with much precision, and this lack of precision can be problematic for clinical and evidence-based purposes. Clinicians use measuring instruments at least in part to provide information regarding whether patients need to change treatment regimens, but to do so they need measurements that provide clear information regarding magnitudes of change. Moreover, one popular use of measuring instruments is to determine efficacy, but this requires the ability to compare magnitudes of change (e.g., between different arms of a study) (see McClimans 2011).

Some of the new psychometric methods, such as those that use the Rasch family of models, claim to generate interval-level measurement. Interval-level measurement allows us to say that a change score of, for example, four, is of the same magnitude no matter where on the scale this change occurs (e.g., whether the change is from 15 to 19 or from 41 to 45). Thus, interval-level measurement allows us to compare change scores across the scale. It also allows us to use mathematical operations, such as the addition or subtraction of a constant, to the measurement values without changing the form of the scale. We can also use the arithmetic mean as a measure of average value. The practical result is that clinicians need not wonder whether a four-point change at one point on the scale represents a different magnitude than the same change at another point in the scale, and researchers can compare magnitudes of change across study populations.

No doubt interval-level scales have practical advantages over ordinal-level scales.[3] But those who advocate for interval-level scales tend to go beyond these practical concerns. One assumption that motivates much of the emphasis on interval-level scales is that ordinal scales are not *scientific measurements*. Measurement, in this view, requires a quantitative variable. This position echoes that of Michell, which I discussed above. It is not an uncommon position. In the paper by Hobart and colleagues (2007), the subheading that leads the section arguing for interval-level measurement is titled "The Requirement for Rating Scales to Generate Rigorous Measurements" and then later, "Ordered Scores [from CTT] Are Not Scientific Measurements."

This emphasis matters. If you see interval-level scales in psychological measurement as having mainly, practical benefits, then the choice of what instrument to use (i.e., instruments with interval-level or ordinal-level scales) will depend on context. But if you understand interval-level measurement as the only form of legitimate scientific measurement, then there is no decision to make: all measuring instruments should be interval-level scales. The risk one wages when using CTT is an epistemic risk to scientific credibility. Proponents of realism in measurement understand their commitment as a commitment to truth and see CTT as committed to, at best, expediency.

To be sure, Borsboom does not hold the same realist view in that he is not committed to interval-level scales. Instead, he sees the commitment to realism as a way of understanding the relation between an attribute and the scores generated by a measuring instrument. One furthers this understanding by choosing a measurement model that explains this relationship. In choosing a measurement model, one must specify the structure of the attribute and the function that relates it to the measurement scores. This process furthers scientific progress at least in part because it provides an argument for the validity of a measuring instrument, and for Borsboom (Borsboom and Zand Scholten 2008), knowing what one is measuring is essential to measuring it.

Validity

Determining validity is the second methodological limitation of CTT that Hobart et al. (2007) discuss in their article. Construct validity, determining whether a measuring instrument measures what it aims to measure, is CTT's primary validation method. It is typically tested by assessing a measure's internal and external construct validity. Internal construct validity is tested by examining the extent to which the questions or items within a measurement scale are statistically related to one another based on the responses given by a sample population. The criticism is that this process does not tell us anything about the construct itself (e.g., quality of life); it tells us only that certain questions tend to behave similarly in the same conceptual space. External construct validity is examined via convergent and divergent validity testing. Multiple measurement scales deemed like and different from one another are applied to a sample population, and the scores derived from respondent answers are correlated. These correlations provide information about whether the scale being validated correlates higher with scales that measure similar constructs than with those measuring dissimilar constructs. Once again, this process does not tell us what construct a measure assesses. It tells us only that some scales are correlated (or not) with other scales.

Construct validity is accused of circularity, vacuity and meaninglessness (Borsboom, Mellenbergh, and van Heerden 2004; Hobart et al. 2007). Hobart and colleagues (2007) argue similarly to Borsboom et al. (2004) that the solution to construct validity is the development of models (Hobart and colleagues refer to them as "construct specification equations") that predict the variation of the data from the measuring instrument. Borsboom and colleagues (2004) add that validity is achieved if and only if (1) the attribute exists and (2) variations in the attribute causally produce variations in the data. For Hobart and colleagues, the idea that the attribute is real is presupposed given that they require interval-level measurement of quantitative variables.

The risk of using CTT in the context of validity is primarily a form of inductive risk. The worry is that claiming validity using tests of internal and external construct validation does not provide sufficient evidence that one is measuring what one intends to measure. Treating statistical correlations as though they provide evidence of an attribute is potentially a bad inference because it is possible that such correlations speak only to the questionnaire as an artifact (i.e., the inference is uncertain). What does the aversion to taking correlations as evidence of an attribute tell us about the values realists prioritize?

At least in part, this aversion speaks to Michell's (2005) and Borsboom's (2005) interest in measurement as furthering scientific discovery. Construct validation as used by CTT purports to provide evidence that an instrument is measuring the attribute it is intended to measure. Insofar as the instrument is validated it also provides evidence of that attribute through its measurement outcomes. If the inference from correlations to an attribute is invalid, then the outcome information that the instrument provides is questionable. If we, nonetheless, take these outcomes as evidence of an attribute, we adulterate the integrity of science and stall progress. To be sure, Borsboom also argues that the use of CTT-based instruments violates our social obligation to the public. But this obligation rests on the assumption that measures that are fit for purpose must be anchored to real attributes. The use of CTT measures is a violation of our social obligation because they are scientifically inadequate, and thus the instruments are materially dishonest.

From a realist perspective, CTT could hardly be worse. On one hand, it purports to measure attributes, and on the other hand, its methods glide across the surface of reality mutually reinforcing the legitimacy of what is a constructed space. Because the attributes are never modeled, nor defined through a robust theory, it is unclear what CTT instruments measure. Yet volumes of articles are devoted to their respective validity. Similarly, because they make use of ordinal scales, it is often unclear how to interpret changes over time, and yet they are used in clinical practice and health policy as evidence of change.

NONEPISTEMIC VALUES AND PSYCHOLOGICAL MEASUREMENT

CTT measuring instruments are easy targets for critics, and I have also criticized them, particularly those CTT instruments used in health care. I (McClimans 2010, 2011) have argued like those above that these measures are invalid and difficult to interpret, and like Borsboom (Borsboom and Zand Scholten 2008), I have argued that these problems ultimately stem from a lack of theorizing about the target attribute.

But unlike the authors above, my concern with these instruments is not limited to a concern with scientific integrity and progress; I am not simply concerned with epistemic values of rigor and truth. Rather, I situate COAs as bearing a double burden of epistemic and ethical credibility. Moreover, because the ethical risk involved in COAs is intertwined with the epistemic choices of measurement methodology and measuring instrument, it is not a matter of one interest taking priority over the other. Thus, we cannot say that a measuring instrument fulfills our social obligation simply because it embodies a certain degree of scientific rigor. Ethical risks must be considered and balanced alongside epistemic risks. Ethical considerations are not new to medical research, but they are relatively absent from discussions of measurement.

Although I think ethical risk is part of many aspects of epistemic practice in measurement, patient-reported forms of COAs present a particularly salient case for taking ethical risk and ethical values seriously. Patient-reported outcome measures (PROMs) measure attributes such as mobility, health status, and quality of life by asking patients questions (e.g., "Does your health now limit you in lifting or carrying groceries?"). PROMs are very popular with health policy makers both in the United States and abroad because they incorporate the patients' point of view into assessments of effectiveness, thus bringing together patient centeredness and clinical effectiveness in one instrument (Black 2013; Department of Health 2008; Speight and Barendse 2010; Washington and Lipstein 2011).

Given their dual purpose, PROMs embody certain ethical-epistemic imperatives. For instance, validity, ostensibly an epistemic concern of measurement, takes on a distinctive ethical dimension when we consider that PROMs are intended to assess the subjective experience of patients' health and well-being by asking them questions about it (Schwartz and Rapkin 2004). If our knowledge of the target attribute is underdeveloped or systematically excludes certain legitimate kinds of subjective experience, then it is not simply that we come to know less about health and well-being, but also we do not live up to what we owe the patients to whom we pose our questions. We may, for example, attribute to them a level of subjective experience they do not have, or we may attribute an experience that is not theirs while making the claim that this is what patients report (and giving ourselves credit for asking them). The nature of PROMs (i.e., that they speak on behalf of patients) means that more is at stake than scientific integrity when we develop and use them (McClimans et al. 2017).

To make my point, I provide two examples. Both examples take up again the concern regarding interpretability and scale development.

Classical Test Theory Reconsidered

Earlier I discussed how CTT measuring instruments are difficult to interpret because they yield ordinal-level scales. Not surprisingly, this difficulty has led to the development of methods to enhance their interpretability. One popular method is the identification of a minimal important difference (MID). A MID is the smallest change in respondent scores that represents clinical significance and which would ceteris paribus warrant a change in a patient's care (Jaeschke, Singer, and Guyatt 1989). One popular method for determining a measure's MID is to map changes in respondent outcomes onto a control. These are referred to as "anchor-based" approaches. The idea is to determine the minimal amount of change that is noticeable to patients and to use this unit of change as the MID.

Here is how it works. A control group of patients are asked to rate the extent of their symptom change over the course of an illness or intervention on a transition-rating index (TRI). TRIs are standardized questionnaires that ask patients questions such as "Do you have more or less pain since your first radiotherapy treatment?" Typically, patients are given seven possible answers ranging from "no change" to "a great deal better" (Fayers and Machin 2015). Those who indicate minimal change (i.e., those who rate themselves as just "a little better" than before the intervention) become the patient control group. The mean-change score of this group is used as the MID for the PROM.

The approach of acquiring a MID via a patient control group assumes that respondents who rate their symptom change as "a little better" on a transition question should ceteris paribus also have comparable change scores from the PROM. Put differently, similarities in respondent answers to transition questions ought to underwrite similarities in respondents' magnitude of change over the course of an intervention or illness. But qualitative data from interviews with patients suggest that this assumption is ill founded (Taminiau-Bloem et al. 2011; Wyrwich and Tardino 2006). To take a concrete example, consider Cynthia Chauhan, a patient advocate during the deliberations on the FDA guidelines for the use of PROMs in labeling claims. In response to the deliberations, Chauhan cautioned those present "not to lose the whole person in your quest to give patient-reported outcomes free-standing autonomy" (Chauhan 2007). To make her point, she discussed the side effects of a drug called bimatoprost, which she uses to forestall blindness from glaucoma. One of the side effects of bimatoprost is to turn blue eyes brown. Chauhan has "sapphire blue" eyes, in which, she says, she has taken some pride. As she speaks of her decision to take the drug despite its consequences, she notes that doing so will affect her identity in that she will soon no longer be the sort of person she has always enjoyed being (i.e., she will no longer have blue eyes). Moreover, she points out that although the meaning that taking this drug has for her is not quantified on any outcome measure, it nonetheless affects her quality of life (Chauhan 2007).

We can imagine that even if the bimatoprost is only minimally successful and Chauhan's resulting change score from the PROM is low, she will nonetheless have experienced a significant change—she will not be the same person she was before. But this significance is tied to the place that her blue eyes had to her identity and what she took to be a good life; ceteris paribus we would not expect a brown-eyed person to summarize his or her experience in the same way. Thus, it would not be surprising if Chauhan's answer to the transition question was "quite a bit" while the magnitude of her change score was minimal.

This example illustrates how a popular and widely used attempt to improve the epistemic quality of CTT instruments is infused with epistemic-ethical concerns. The assumption that motivates the use of a MID is ill founded. Patients' assessments of change are influenced by the way that change interfaces with their identity and their accounts of what makes for a good quality life. Using a MID as the interpretive key for a PROM threatens to lose the "whole person," as Chauhan put it, while only providing the appearance of epistemic improvement. Epistemically, MIDs do not necessarily underwrite any particular magnitude of change from PROMs; thus, they do not necessarily represent the smallest clinically significant change. Using a MID as though it represents clinical significance threatens to ignore or exacerbate the harms or benefits patients experience depending on the particulars of the intervention in question.

Latent Trait Theory Reconsidered

Earlier I discussed that, the Rasch family of models claim to be able to establish interval-level measurement. For my purposes here, I'm going to assume that in principle this claim is true (for a debate over this claim see Borsboom and Mellenbergh 2004; Borsboom and Zand Scholten 2008; Michell 2000).

To provide some background, the Rasch model requires (1) that a person with greater ability should have a greater probability of answering a question correctly and (2) that given two questions, one of which is more difficult than the other, a person has a greater probability of answering the easier question. Given these relationships, Rasch models can establish measurement values for person ability and question difficulty. The function used in Rasch models is a logit, and the formula for one variant of this model for questions with dichotomous response options is:

$$Pr\{x_{ni}|\beta_n,\delta_i\} = \frac{e^{x_{ni}(\beta_n-\delta_i)}}{1+e^{(\beta_n-\delta_i)}}$$

where $x_{ni} \in [0,1]$; β_n and δ_i are the measurements of the ability of person n and the difficulty of item i, respectively, upon the same latent trait, and e is the natural logarithm constant (2.718). In COA applications, these models often concern levels of severity or frequency on a rating scale interpreted as indicating less or more health, disease burden, functionality, engagement in decision making, and so on.

The extent to which observed data (patients' responses to questions) "fit" the predictions of those responses from a Rasch model, within acceptable uncertainty, indicates the extent to which "measurement" is achieved. This is because, if ordered items fit the Rasch model's predictions, we can infer interval-level measurement of a latent variable. But we know that observed data never perfectly fit the predictions of a model. Thus, where do we set the threshold? To be sure, we are looking for a convergence between the empirical data and the Rasch model that is able to support useful inferences, but whether the convergence is sufficient for such inferences is open to legitimate disagreement.

This area of legitimate disagreement is a place where nonepistemic values come into play. Consider a questionnaire whose data imperfectly fit the Rasch model, but the way they imperfectly fit the model has an easy "fix": you can get rid of some of the questions in the original questionnaire.[4] Should you remove these questions to get a better fit? The qualitative research during the development of the questionnaire suggested these questions were important—research that involved patients and clinical experts in the field—but if we get rid of them, we can claim that this attribute is a continuous quantity. Doing so might mean that we can meet the demands of the government agency or pharmaceutical company funding us; it also might make us look good, and/or we might be able to sell the instrument and make a profit. But we might also consider how removing these items could affect how people fare when the attribute in question is measured by the instrument. Ethically speaking, might removing the items harm people; could it benefit them? Moreover, although we can claim that the attribute is a continuous quantity, removing the items weakens this claim and thus weakens the argument of an epistemic advance of Rasch over CTT.

CONCLUSION

Contemporary debate on psychological measuring instruments tends to focus on the importance of a realist ontology. I have argued that proponents of this position understand realism to support epistemic values of truth and rigor in measurement, values they take to be undermined by using antirealist approaches to measurement, namely CTT. I have suggested that whatever the inadequacies of CTT, these problems cannot be considered purely ontological or epistemic failings, and neither can the measures from latent trait theory be understood as simply advancing science. Psychological measures, particularly PROMs, are characterized by ethical-epistemic entanglements. In some cases, these entanglements mitigate against realist claims as they do in the example of Rasch, or they can complicate the antirealist picture psychometricians attribute to CTT as they do in the example of MIDs.

What, I think, these entanglements illustrate is that the emphasis on a realist ontology as a threshold over which serious psychologists and legitimate measures must cross is overly simplistic. Nonepistemic, often ethical values, enter into a large proportion of questions during measurement development and use—both for CTT and latent trait theory. These questions are not divisible from the epistemic questions whose values realists seek to uphold and whose answers are often used as evidence of a realist ontology. Ignoring them falsely suggests that latent trait theory is inherently superior to CTT both scientifically and in terms of our obligations to respondents.

NOTES

1. Borsboom (2005) argues for a between-subject causal connection between latent variables and item responses (opposed to within-subject causal account, which he rejects).

2. To be sure, operationalism need not imply antirealism. As Eran Tal (2016) points out, the methods chosen could underwrite an attribute that refers to a mind-independent reality. But the way CTT is used tends to render this interpretation irrelevant since the point Borsboom and Michell are making is that psychological measurement requires realism to be a form of measurement.

3. I have argued that in some cases they have some epistemic advantages as well; see McClimans, Browne, and Cano, 2017.

4. This is not simply a thought experiment. See, for example, the development of the ABILHAND questionnaire, where items that failed the statistical tests of fit to the Rasch model were discarded (Durez et al. 2007).

REFERENCES

Black, Nick. 2013. "Patient Reported Outcome Measures Could Help Transform Healthcare." *The British Medical Journal* 346: f167. doi:10.1136/bmj.f167.

Borsboom, Denny. 2005. *Measuring the Mind*. Cambridge: Cambridge University Press.

Borsboom, Denny. 2006. "The Attack of the Psychometricians." *Psychometrika* 71: 425–40. doi:10.1007/s11336-006-1447-6.

Borsboom, Denny, and Gideon J. Mellenbergh. 2004. "Why Psychometrics Is Not Pathological: A Comment on Michell." *Theory & Psychology* 14: 105–20. doi:10.1177/0959354304040200.

Borsboom, Denny, Gideon J. Mellenbergh, and Jaap van Heerden. 2004. "The Concept of Validity." *Psychological Review* 111: 1061–71. doi:10.1037/0033-295X.111.4.1061.

Borsboom, Denny, and Annemarie Zand Scholten. 2008. "The Rasch Model and Conjoint Measurement Theory from the Perspective of Psychometrics." *Theory & Psychology* 18: 111–17. doi:10.1177/0959354307086925.

Bridgman, Percy. 1927. *The Logic of Modern Physics*. London: The Macmillan Company.

Cano, Stefan J., and Jeremy C. Hobart. 2011. "The Problem with Health Measurement." *Patient Preference and Adherence* 5: 279–90. doi:10.2147/PPA.S14399.

Chauhan, Cynthia. 2007. "Denouement: A Patient-Reported Observation." *Value in Health* 10: S146–S47. doi:10.1111/j.1524-4733.2007.00276.x.

Department of Health. 2008. *High Quality Care for All: NHS Next Stage Review Final Report*. London: The Stationary Office.

Durez, Patrick, Virginie Fraselle, Frédéric Houssiau, Jean-Louis Thonnard, Henri Nielens, and Massimo Penta. 2007. "Validation of the ABILHAND Questionnaire as a Measure of Manual Ability in Patients with Rheumatoid Arthritis." *Annals of the Rheumatic Diseases* 66: 1098–105. doi:10.1136/ard.2006.056150.

Fayers, Peter, and David Machin. 2015. *Quality of Life: The Assessment, Analysis and Interpretation of Patient-Reported Outcomes*. 2nd Edition. Hoboken, NJ: John Wiley & Sons.

Heilmann, Conrad. 2015. "A New Interpretation of the Representational Theory of Measurement." *Philosophy of Science* 82: 787–97. doi:10.1086/683280.

Hobart, Jeremy, and Stefan Cano. 2009. "Improving the Evaluation of Therapeutic Interventions in Multiple Sclerosis: The Role of New Psychometric Methods." *Health Technology Assessment* 13: 1–177. doi:10.3310/hta13120.

Hobart, Jeremy C., Stefan J. Cano, John P. Zajicek, and Alan J. Thompson. 2007. "Rating Scales as Outcome Measures for Clinical Trials in Neurology: Problems, Solutions, and Recommendations." *Lancet Neurology* 6: 1094–105. doi:10.1016/S1474-4422(07)70290-9.

Jaeschke, Roman, Joel Singer, and Gordan H. Guyatt. 1989. "Measurement of Health Status: Ascertaining the Minimal Clinically Important Difference." *Controlled Clinical Trials* 10: 407–15.

Lord, Frederic M., and Melvin R. Novick. 2008. *Statistical Theories of Mental Test Scores*. Charlotte: Information Age Publishing.

Maul, Andrew. 2013. "On the Ontology of Psychological Attributes." *Theory & Psychology* 23: 752–69.

McClimans, Leah. 2010. A Theoretical Framework for Patient-Reported Outcome Measures. *Theoretical Medicine and Bioethics* 32: 47–60.

McClimans, Leah. 2011. "Interpretability, Validity, and the Minimum Important Difference." *Theoretical Medicine and Bioethics* 32: 389–401. doi:10.1007/s11017-011-9186-9.

McClimans, L, Browne, J and Stefan, C. 2017. "Clinical Outcome Measurement: Models, Theory, Psychometrics and Practice", *Studies in the History and Philosophy of Science* (DOI) 10.1016/j.shpsa.2017.06.004 available at https://doi.org/10.1016/j.shpsa.2017.06.004

Michell, Joel. 1997. "Quantitative Science and the Definition of Measurement in Psychology." *British Journal of Psychology* 88: 355–83.

Michell, Joel. 1999. *Measurement in Psychology: A Critical History of a Methodological Concept*. Cambridge: Cambridge University Press.

Michell, Joel. 2000. "Normal Science, Pathological Science and Psychometrics." *Theory & Psychology* 10: 639–67. doi:10.1177/0959354300105004.

Michell, Joel. 2005. "The Logic of Measurement: A Realist Overview." *Measurement* 38: 285–94.

Schwartz, Carolyn, and Bruce Rapkin. 2004. "Reconsidering the Psychometrics of Quality of Life Assessment in Light of Response Shift and Appraisal." *Health and Quality of Life Outcomes* 2: 16.

Speight, Jane, and Shalleen M. Barendse. 2010. "FDA Guidance on Patient Reported Outcomes." *British Medical Journal* 340: c2921. doi:10.1136/bmj.c2921.

Tal, Eran. 2013. "Old and New Problems in Philosophy of Measurement." *Philosophy Compass* 8: 1159–73. doi:10.1111/phc3.12089.

Tal, Eran. 2016. "Measurement in Science." In *The Stanford Encyclopedia of Philosophy* (Winter 2016 Edition), edited by Edward N. Zalta. https://plato.stanford.edu/archives/win2016/entries/measurement-science/.

Taminiau-Bloem, Elsbeth F., Florence J. Van Zuuren, Mechteld R. M. Visser, Carol Tishelman, Carolyn E. Schwartz, Margot A. Koeneman, Caro C. E. Koning, and Mirjam A. G. Sprangers. 2011. "Opening the Black Box of Cancer Patients' Quality-of-Life Change Assessments: A Think-Aloud Study Examining the Cognitive Processes Underlying Responses to Transition Items." *Psychology & Health* 26: 1414–28. doi:10.1080/08870446.2011.596203.

Washington, A. Eugene, and Steven H. Lipstein. 2011. "The Patient-Centered Outcomes Research Institute: Better Information, Decisions, and Health." *New England Journal of Medicine* 365: e31.

Wyrwich, Kathleen W., and Vicki M. Tardino. 2006. "Understanding Global Transition Assessments." *Quality of Life Research: An International Journal of Quality of Life Aspects of Treatment, Care and Rehabilitation* 15: 995–1004. doi:10.1007/s11136-006-0050-8.

Chapter 7

The Epistemological Roles of Models in Health Science Measurement

Laura M. Cupples

Patient-reported outcome measures are survey instruments used by health researchers and clinicians to quantify health-related quality of life or health status.[1] These measures are epistemically sound only when they can be shown to be valid, comparable to other measures of the same attribute, and accurate. In this paper, I introduce three different kinds of models that I argue are essential for supporting judgments of validity, comparability, and accuracy, respectively. The first types of models are qualitative models. These models represent patients' and researchers' interpretations of test items and their conceptualizations of target attributes. Second, I examine statistical models; they are models that give an account of how patients interact with questionnaire items. The third kind of models I discuss are theoretical models. These models tell a story about the composition of the attribute, its behavior over time and across patient groups, and the relationship between patients' raw scores and the level of the attribute they possess.

While other authors have discussed the roles of qualitative models (McClimans 2010), statistical models (Bond and Fox 2007; Streiner, Norman, and Cairney 2015), and theoretical models (Rapkin and Schwartz 2004; Stenner et al. 2013), in many cases they have not tied these models to their epistemic roles. That is, they have not necessarily associated them with judgments about content validity, comparability, and accuracy.

BACKGROUND

In what follows, I discuss the relationship between patient-reported outcome measures and the models that I contend ought to be used to support them. Patient-reported outcome measures are survey instruments used by medical researchers and clinicians to quantify patients' health status or health-related quality of life. These measures rely on self-reports to make patients' private experiences public and accessible to clinicians and researchers. They typically ask respondents questions about, for example, physical and psychological functioning, mobility, social connectedness, pain levels, or other factors that researchers believe contribute to health status and health-related quality of life.

While some of the instruments used to measure health status and health-related quality of life are generic (e.g., the Short Form-36 [Stewart and Ware 1992] and the Nottingham Health Profile [McDowell 2006]) and thus supposed to quantify well-being for patients with a wide range of ailments and health statuses, other instruments are disease specific and designed to be used with patients who have only particular illnesses (e.g., asthma, arthritis, or cancer). Still other instruments are site specific and focus on the effect of injury to, deterioration of, or intervention upon certain body parts (e.g., the Oxford Hip Score [Murray et al. 2007] and the BREAST-Q [Klassen et al. 2009]).

Measurement of health-related quality of life and health status involves complex processes, which include patient understandings and interpretations of survey questions, the cognitive abilities of patients, their powers of memory, and the values that shape their appraisal of quality of life. Moreover, patient interactions with survey items also depend on the statistical properties of those items and their intended conceptual content. Furthermore, measurement involves the numeric representation of outcomes and the management of error. Models of the measurement process are holistic representations that consider some subset of these factors. I will argue below that to obtain a full picture of the measurement process and to facilitate judgments about an instrument's validity, accuracy, and comparability, three different models of the measurement process must be deployed, namely qualitative models, statistical models, and theoretical models.

Throughout this chapter, I will understand models to be abstract and idealized representations of dynamic systems that are constructed based on theoretical, statistical, and pragmatic assumptions about those systems. While models are based in part on abstract theory, they also function separately from that theory because they often incorporate material constraints and affordances, assume background conditions specific to the local system in question, and reflect the limits of our mathematical capabilities (Morgan and Morrison 1999).

QUALITATIVE MODELS AND CONTENT VALIDITY

In this section, I argue that qualitative models of the measurement process have an important role to play in supporting judgments about the content validity of measures in the health sciences. I take a qualitative model of the measurement process to be an explication of patient or researcher interpretations of test items. These interpretations help to determine the actual conceptual content of the measure since they affect the operationalization of the measure. Yet we also hope that the intended conceptual content of the measure matches up with patient conceptualizations and interpretations. Unfortunately, patients and researchers often understand test items and target attributes, such as health status and health-related quality of life, differently from one another and differently over time (McClimans 2010; Rapkin and Schwartz 2004). Varying understandings of the attribute in question mean that patients can interpret the meanings of test questions in different ways. Thus, patients may, in effect, be answering different questions from the ones researchers believe themselves to be asking. As I will explain below, when this happens, the content validity of our measures suffers.

A measure with good content validity comprehensively covers all domains that are part of the target attribute. All and only those domains that are part of the target attribute are captured by such a measure (Food and Drug Administration 2009). Content validity is important because it helps to secure inferences from a measure's outcomes to an attribute of interest (i.e., that the quantitative representation given by the measure's outcome is representative of some portion or level of the attribute). If a measure is intended to assess quality of life after mastectomy and breast reconstruction but the items focus on physical functioning and neglect aesthetic appearance, then we might reasonably lack confidence in the inference that the measure's outcome represents quality of life after these interventions. Our lack of confidence is because the measure has poor content validity (i.e., it neglects aspects of the attribute that are relevant to making inferences from the outcomes).

But how is content validity diminished by a mismatch in patients' and researchers' interpretations of test items? When patients' interpretations of test items fail to coincide with the interpretations envisioned by quality-of-life researchers, the operationalization of the measure when applied to patient populations may differ from the operationalization intended by researchers. Test items will carry different meanings, and thus different conceptual content, from what was envisioned. This difference results in diminished content validity because the inference from the measure's outcomes to the intended attribute is invalid. The instrument does not, in fact, measure what researchers meant for it to measure.

Why do patients and researchers sometimes disagree in their understandings and interpretations of items? Moreover, what might such disagreement look like? Imagine we are trying to get a sense of how limited patients are in their mobility. To determine this, we ask several groups of patients how difficult it is for them to engage in strenuous exercise. Healthy patients may envision a run of several kilometers, while for patients with a chronic illness or disability, a walk of a few hundred meters may be considered strenuous. For very elderly patients or patients with significant disability, even a walk across the house may be challenging. Because of the different contexts informing their interpretations, these patients are answering different test items from one another and perhaps different test items from those researchers may have intended them to answer. Depending on the contrast class they apply to the question (for instance, how limited in their mobility they were a month ago, how limited they perceive other patients with the same illness or injury might be, or how limited they were when in full health [see van Fraassen 1980; McClimans 2011]), patients may see a broad range of abilities as indicative of relatively good mobility for them (Rapkin and Schwartz 2004). When they talk about how limited they are in their mobility, even patients who cannot engage in very strenuous activity may feel less limited than we might imagine. On the flip side, patients who are still relatively mobile may feel more limited than we might see them as being.

If we want our measures to demonstrate good content validity, we need to find a way to bring the qualitative models of the measurement process—the models that specify patients' and researchers' interpretations of test items and therefore the conceptual content of the measures we are interested in—into agreement with one another. How can we best accomplish this goal? Because patient-reported outcome measures were created to give patients a voice regarding their own subjective health status, it seems that we should privilege their understandings of health status and health-related quality of life. This means that researchers should concentrate on building qualitative models that describe patients' true interpretations of test items. Researchers cannot simply assume that mismatches between their interpretations and those of patients constitute error on the part of patients. They must reexamine their own interpretations considering those held by patients (McClimans 2010).

How can researchers discover the content of patients' conceptualizations of health status and health-related quality of life, and how can they learn about patients' interpretations of test items? This is done through qualitative research during the instrument development process. Patient focus groups are asked about the domains they feel are most important to their health-related quality of life or health status. They can be asked which symptoms make the biggest impact on their lives and which capabilities are most important for them to maintain. This sort of information, along with input from clinical experts, helps researchers write items that are relevant to patients' experiences with health and illness (Klassen et al. 2009). Once a draft of the instrument has been completed, patients can be interviewed individually as part of a think-aloud study (Bellan 2005; Westerman et al. 2008). Patients can be queried about the relevance and clarity of test items (i.e., about how they interpret the items they are presented with and why). When researchers have access to these interpretations and can write instruments that cover the conceptual content that patients feel is most relevant to the attributes to be measured, they will be able to build relatively accurate qualitative models of the measurement process.

The good news is that most quality-of-life researchers now rely on patient input during measure development. The practice of interviewing patients about their experiences with illness and treatment has become much more common since the 2009 publication of a new FDA guidance on the development of patient-reported outcome measures. This recent change in practice is an important first step in establishing sound qualitative models.

STATISTICAL MODELS AND COMPARABILITY

In this section I discuss two statistical models used to represent the process of health status and health-related quality-of-life measurement. Specifically, I will examine the model(s) used in classical test theory (CTT) and those used by Rasch measurement theory. In general, the models used by CTT give an account of how observed scores relate to true scores (Streiner et al. 2015), and the models used by Rasch represent how patients interact with test items to produce an outcome or test score (Stenner et al. 2013). In what follows, I examine the ways these statistical models epistemically support or fail to support judgments about comparability among measures of the same attribute.

Classical Test Theory

Most patient-reported outcome measures are designed and analyzed using CTT. While modern psychometric methodologies, such as Rasch measurement theory, boast greater utility in many respects (e.g., CTT produces ordinal-level measures, and Rasch produces interval-level measures), CTT is still very popular due to its flexibility and ease of use. CTT employs a thinner statistical model than modern psychometric theories, such as Rasch. Because of the way it models measurement, it gives us little information about the mechanics of the measurement process or about the ways patients interact with individual test items (Borsboom 2005). Moreover, as I will show, the model employed by CTT does not easily facilitate the creation of comparable measures of the same target attribute (Bond and Fox 2007).

The CTT model posits three variables: a true score (T_T), an observed score (T_O), and a random error term (E):

$$T_O = T_T + E$$

We can think of a respondent's true score as the expected value of the observed score (the actual score achieved on the measure) over a universe of possible observations of the same construct. As shown above, the observed score is the sum of the true score plus the random error term. The expected value of the random error term over many test administrations is zero (Borsboom 2005).

In CTT, the individual items are taken to be members of a random sample drawn from a population of possible items (Kane 1982). Answers to each item contribute equally to the final raw score, and what matters is how items perform en masse rather than individually (Streiner et al. 2015). This is because the unit of analysis in CTT is the test rather than, say, individual items in the questionnaire. The result is that CTT gives us little or no insight into how respondents interact with individual test items. For example, CTT does not specify the difficulty of each test item, nor does it tell us how likely it is that a respondent with a certain level of the target attribute will answer an item in a particular way.[2] Instead of providing information at the item level, CTT helps us understand how groups of respondents interact with the test as a whole. In what is called norm-referenced measurement, patients' scores on CTT tests are compared with the performance of norm groups to place outcomes in context.

Because CTT focuses on how groups of respondents interact with the test, it is difficult to achieve comparability of measuring instruments. In other words, it is difficult to develop parallel measures of the same attribute for which the same scores carry the same meaning (i.e., signify the same level of quality of life or health status). To say that two instruments measure the same attribute, we must ensure that test items cover the same range of content. But this coverage is difficult to ensure with CTT at least in part because attributes measured by CTT instruments are often multidimensional (Borsboom 2005) (e.g., health status and health-related quality of life are usually taken to include physical, functional, emotional, and social dimensions [Cella 1994]). In a CTT measure, the content of the attribute is determined by the specific content of the totality of the questions (Streiner et al. 2015). It is tricky to perfectly replicate the conceptual content of a CTT test even if you try to match questions by conceptual content item by item. For instance, do turning a key and fastening a button require the same type of capability? Or do questions about these two tasks in fact cover different conceptual content? Nonetheless, with good qualitative and theoretical models of the measurement process, it may be possible to create tests that measure the same attribute. When good conceptual definitions are used to inform the content of test items, there is a better chance that those items will cover the same conceptual content as comparable tests. This is because good conceptual definitions can help answer exactly such questions as whether fastening a button and turning a key require the same sort of capability. However, a common criticism of patient-reported outcome measures is that their conceptual and theoretical foundations are usually rather weak (McClimans 2010; Hobart et al. 2007; Hunt 1997). This means that these sorts of questions are usually left unanswered.

In addition to ensuring that two instruments measure the same attribute, comparability requires that the same scores carry the same meanings on those instruments. CTT tests differ in how their scores are determined. Most tests simply sum responses to questions to arrive at a raw score, but once this is done, they often rescale that raw score in some way to arrive at an outcome. The algorithm used to rescale outcomes differs from instrument to instrument (e.g., the Disabilities of the Arm, Shoulder and Hand in Cano et al. [2011] and the nonnormed physical function scale for the Short Form-36 in Stewart and Ware [1992]). For this reason, identical scores on two different instruments may signify different levels of the same attribute. Similarly, questions may be posed either positively or negatively—targeting, for instance, either mobility or its inverse.[3] For this reason, even the directionality of scales may differ.

Lately, efforts have been made by quality-of-life researchers to place scores on normed scales. By placing outcomes on a scale from 0 to 100, calibrating mean score values to 50, and scaling standard deviations to 10 for several quality-of-life instruments, researchers have been able to facilitate comparability among measures of the same attribute (e.g., the Short Form-36 in Stewart and Ware [1992]). Unfortunately, these efforts can also be misleading. Placing outcomes on the same scale does not ensure that measures cover the same conceptual content and thus does not ensure that they target the same attribute. As I have argued, two requirements must be met for comparability between measures. Measures must target the same attribute, and like scores must carry like meanings.

Rasch Measurement Theory

Recently health researchers have begun to take advantage of the resources offered by modern testing methodologies, such as Rasch measurement theory and item response theory (IRT) (Hobart et al. 2007). The Rasch model is often considered to be a subset of IRT models, so for the sake of simplicity, I will focus on the Rasch model in this section.[4] Rasch measurement theory deploys a thicker statistical model than CTT primarily because it tells a more complete story about how patients with a certain level of the measured attribute interact with individual test items of varying difficulty. Rasch locates an instrument's items on a continuum according to difficulty so that successively ranked items should each be more difficult for patients to answer (Bond and Fox 2007). The more items a patient can endorse on the BREAST-Q, for instance, the more favorable her surgical outcome is estimated to be in terms of satisfaction with surgical results and care as well as resultant quality of life (Klassen et al. 2009).

Unlike CTT Measures, instruments designed and analyzed using Rasch measurement theory are intended to measure unidimensional attributes. The level of attribute possessed by the patient counters the difficulty of individual test items so that when the level of attribute exceeds the difficulty of a test item, a patient is more likely to answer a question in the affirmative (Bond and Fox 2007). For instance, the Patient-Reported Outcome Measures Information System (PROMIS) physical function instrument asks questions such as "Are you able to sit at the edge of your bed?" and "Are you able to carry a laundry basket up a flight of stairs?" Intuitively, a patient must possess more mobility to be able to answer the second question in the affirmative than the first. Rasch measurement theory tells us the probability (P) that an item (x_i) will be answered in a particular way ($P(x_i = 1)$)—for instance, that an item will be endorsed (1) rather than rejected (0) is determined solely by the relationship between item difficulty (d_i) and the amount of the attribute possessed by the patient (b) (Stenner et al. 2013). So, for example, the probability that a patient will agree that she is able to dress herself depends on the relationship between the difficulty of the task and her amount of functional ability. The equation below describes what is called an item response curve, for a given item (x_i). This equation is graphically represented in Figure 7.1. An item response curve describes the probability that an item of a given difficulty will be endorsed or that a particular answer will be given, based on the level of attribute possessed by the respondent.

$$P_r(x_i = 1) = \frac{e^{(b - di)}}{1 + e^{(b - di)}}$$

The Rasch model boasts several advantages over CTT. For instance, because of the mathematical separability of item difficulty and level of attribute in the Rasch model, these two factors are invariant across patient populations and with respect to the subset of test items employed, respectively. That is, the difficulty of items does not depend on who is responding to them or on how much of the measured attribute they possess. Likewise, estimates of a patient's level of the measured attribute do not depend on the specific items employed by the measure. The function of the measuring instrument does not depend on the context in which it is employed (i.e., whether a meter stick is used to measure a table or a rug), and measurement outcomes do not depend on the specific instrument used if that instrument is properly calibrated. Together, these qualities are often referred to as specific objectivity (Stenner and Burdick 1997). The specific objectivity of these types of measures is an extremely useful trait because it makes it possible to compose comparable

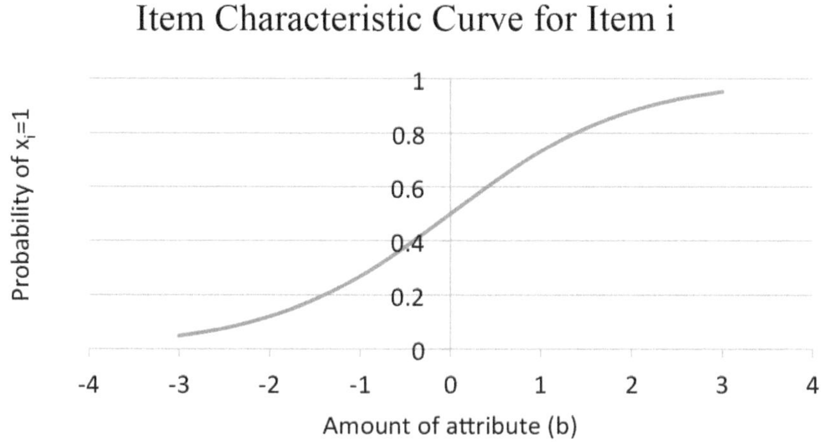

Figure 7.1 Item characteristic curve showing the probability of the respondent choosing the answer $x_i = 1$. The difficulty of item i is set to 0 logits. *Source:* Stenner, A. Jackson, et al. 2013. "Causal Rasch Models." *Frontiers in Psychology* 4: 1–14.

tests of the same attribute using the method of item banking. With item banking, a large bank of items is created, with all items measuring the same unidimensional attribute, and subsets of items from that bank are combined to form tests of various lengths (often with the goal of minimizing the burden placed on patients) (Bond and Fox 2007). Tests can also be created that are targeted to patients with a particular amount of the measured attribute so that the instrument provides more precise measurement at that attribute level.

Nevertheless, it is still important when composing comparable instruments using Rasch to base those measures on qualitative models of the target attribute. Though the mathematical characteristics of Rasch measurement can ensure that these measures are both unidimensional and specifically objective, and hence that the various instruments composed from associated item banks all measure the same attribute, it is still important to know what the conceptual content of that attribute is (i.e., to ensure good content validity). For instance, it is important to know whether a measure targets depression or anxiety. These two attributes are often comorbid, and similar questions can be used to assess them. Thus, a good qualitative model is necessary to separate measures of one from the other.

THEORETICAL MODELS AND ACCURACY

In this section, I discuss the epistemic role of theoretical models of the measurement process and argue that they facilitate judgments about measurement accuracy. Theoretical models of the measurement process are models informed by a theory of the attribute. Like qualitative models, they tell us about the conceptual content of the measure. They might tell us, for instance, when it is permissible to drop a statistically ill-fitting item from a Rasch measure and when that item is essential to the instrument's conceptual content. But they also tell us how the attribute behaves—how it changes over time and across circumstances or patient groups. So a theoretical model would tell us what kind of change in quality of life we might expect over the course of a patient's illness or treatment and whether an unexpected change should be classed as a legitimate variation in the target attribute or an instance of error (McClimans 2010).

Marjan Westerman and her colleagues (2008) studied a phenomenon called response shift among a group of cancer patients receiving chemotherapy. Response shift is an unexpected change in a patient's measured level of quality of life or some other target attribute due to adaptation to illness or treatment. For instance, a patient may change her frame of reference—the standard to which she compares her current condition—and this may alter her appraisal of her quality of life. Or a patient may reconceptualize what it means to be limited in his pursuit of leisure activities. One of Westerman's cancer patients claimed at the beginning of treatment that he was very limited in pursuing his leisure activities. He was an avid gardener, but found his hobby difficult to maintain once he became ill. Several weeks later, he claimed he was only a little bit limited; yet by all accounts, he was more physically limited than when he responded the first time (Westerman et al. 2008, 555). Most quality-of-life researchers hold the pretheoretic assumption that quality of life and the domains that make it up are standardizable. That is, they take the meaning of quality of life, or of limitation in this case, to be constant from one case to the next. Thus, when these concepts shift in their meaning, as they did for Westerman's patient, they assume it must be due to measurement error. But a theory of the measured attribute might suggest that quality of life and its constituent domains cannot be standardized in this way. It may suggest that meanings shift according to patients' circumstances. If so, this patient's reconceptualization of what it means to be limited might not be an instance of measurement error at all but instead an example of legitimate qualitative variation in the target attribute (McClimans 2010; Rapkin and Schwartz 2004).

A theoretical model helps us make judgments about measurement accuracy in part by allowing us to distinguish between legitimate changes in a patient's level of quality of life and responses that should be considered errors. Without a theory of quality of life, it is premature to make the judgment that the patient whose quality of life appeared to improve—due to his adaptive change in leisure activities—was in error about his quality of life (McClimans

2010). Quality of life may in fact change based on subjective assessments of limitation rather than due to objective improvement or deterioration. Many people with acquired disabilities, after initially rating their quality of life as lower than when they were able bodied, later claim to value their new lives just as highly as their previous, able-bodied lives (Barnes 2016). Taking their testimony seriously may require us to see changes in quality of life due to response shift as legitimate variation.

Some proponents of Rasch measurement (Hobart et al. 2007; Stenner et al. 2013) see a somewhat different role for theoretical models of the measurement process. They see these models essentially as helpers to the statistical model. According to Jack Stenner and his colleagues (2013), theoretical models help to predict the difficulty of test items based on certain causal factors that explain their position on the measurement scale. For instance, the difficulty of mobility items might vary in terms of the strength and range of motion required to complete the relevant mobility tasks.[5] When testing these theoretical models, we can compare our empirical estimations of item difficulty (estimations based on the probability distribution of actual patient responses) with our calculated theoretical values for mobility to determine how closely empirical values match theoretically calculated values. Once our theoretical model has been well confirmed, we can use it to make judgments about item fit to the model.

I suggest that in the case of Rasch measures, having this sort of theoretical model of the measurement process allows us to make judgments about what Eran Tal calls operational accuracy—or accuracy relative to some standard (2012). We use measurement standards to calibrate individual instruments, or to ensure that they conform to an idealized scale and thus measure their target attributes accurately. In this case, the standard against which empirical measures of item difficulty are being calibrated is the idealization of item difficulty predicted by the theoretical model.[6] Without a theory of the attribute to facilitate calculation of theoretical values for item difficulty, we do not have a standard for comparison with empirical values, and we cannot make judgments about the operational accuracy of our measures. That is, we cannot make judgments about the accuracy of our measures relative to the standard set by theory.

Unfortunately, patient-reported outcome measures have notoriously weak conceptual grounding and in most cases, lack a theory of the attribute and a forteriori, a theoretical model of the measurement process (Hobart et al. 2007; Hunt 1997; McClimans 2010). This frequent lack of theoretical model has consequences for our ability to make judgments about the accuracy of patient-reported outcome measures for both CTT and Rasch measures. Thus, one suggestion for future research is to develop good theoretical models for these measures. Not only will uncovering these models aid us in making judgments about measurement accuracy, but since theoretical models also incorporate certain roles of qualitative models, they will also aid us in making judgments about content validity.

CONCLUSION

The model-based account of measurement epistemology developed by Eran Tal (2012) for use in physical measurement argues that for us to make legitimate inferences about measure validity, comparability, and accuracy, our measures must be epistemically supported by abstract and idealized models of the measurement process. I have discussed three broad types of models of the measurement process for patient-reported outcome measures of health-related quality of life and health status. Qualitative models reflect patients' understandings and interpretations of the construct in question and the associated test items. These models facilitate judgments about content validity. Statistical models give an account of how patients interact with test items in the Rasch framework and how observed scores relate to true scores in the CTT framework. The statistical model that a measure is rooted in helps to determine how comparable measures of the same attribute can be constructed. Finally, theoretical models are models that are derived from a theory of the measured attribute. Not only do they tell us about the conceptual content of the measure; they also tell us about the behavior of the target attribute over time and across patient populations. Theoretical models help us distinguish legitimate variation in the target attribute from patient error, and they help us establish the operational accuracy of Rasch measures.

NOTES

1. I would like to thank Leah McClimans for helpful comments on earlier versions of this chapter.

2. Indeed, though I use the language of attributes for the sake of consistency, the CTT framework (unlike the Rasch framework) need not even hypothesize the existence of an underlying causal attribute. In general, CTT speaks of constructs rather than attributes.

3. It is debatable whether positively and negatively worded questions about, say, mobility even measure the same attribute. See, for instance, Anatchkova, Ware, and Bjorner (2011).

4. The mathematical model deployed by Rasch measurement theory is identical to the one-parameter item response theory model. The only distinction is that the

Rasch model is prescriptive, while the item response model aims only to be descriptively accurate (Andrich 2004). Two- and three-parameter IRT models incorporate additional variables to better describe measurement data, but in doing so, they forfeit certain functional advantages shared by Rasch and the one-parameter model.

5. A. Jackson Stenner, telephone conversation with author, June 1, 2016.

6. In his 2012 article, "How Accurate Is the Standard Second," Tal notes that the duration of the standard second is defined and determined by idealized models, not by the ticks of physical clocks. This is because the duration of the tick of even the best physical clock is disrupted by several outside forces that carry it away from the duration described by the definition.

REFERENCES

Anatchkova, Milena D., John E. Ware Jr., and Jakob B. Bjorner. 2011. "Assessing the Factor Structure of a Role Functioning Item Bank." *Quality of Life Research* 20: 745–58.

Andrich, David. 2004. "Controversy and the Rasch Model." *Medical Care* 42: I-7–I-16.

Barnes, Elizabeth. 2016. *The Minority Body: A Theory of Disability*. Oxford: Oxford University Press.

Bellan, Lorne. 2005. "Why Are Patients with No Visual Symptoms on Cataract Waiting Lists?" *Canadian Journal of Ophthalmology* 40: 433–38.

Bond, Trevor G., and Christine M. Fox. 2007. *Applying the Rasch Model: Fundamental Measurement in the Human Sciences*. 2nd Edition. New York: Routledge, Taylor & Francis Group.

Borsboom, Denny. 2005. *Measuring the Mind: Conceptual Issues in Contemporary Psychometrics*. Cambridge: Cambridge University Press.

Cano, Stefan, Louise E. Barrett, John P. Zajicek, and Jeremy Hobart. 2011. "Beyond the Reach of Traditional Analyses: Using Rasch to Evaluate the DASH in People with Multiple Sclerosis." *Multiple Sclerosis Journal* 17: 214–22.

Cella, David F. 1994. "Quality of Life: Concepts and Definition." *Journal of Pain and Symptom Management* 9(3): 186–92.

Food and Drug Administration. 2009. *Guidance for Industry on Patient-Reported Outcome Measures: Use in Medicinal Product Development to Support Labeling Claims*. Federal Register 74: 1–43.

Hobart, Jeremy, and Stefan Cano. 2009. "Improving the Evaluation of Therapeutic Interventions in Multiple Sclerosis: The Role of New Psychometric Methods." *Health Technology Assessment* 13: 1–177.

Hobart, Jeremy, Stefan J. Cano, John P. Zajicek, and Allan J. Thompson. 2007. "Rating Scales as Outcome Measures for Clinical Trials in Neurology: Problems, Solutions, and Recommendations." *Lancet Neurology* 6: 1094–105.

Hunt, S. M. 1997. "The Problem of Quality of Life." *Quality of Life Research*, 6: 205–12.

Kane, Michael T. 1982. "A Sampling Framework for Validity." *Applied Psychological Measurement* 6: 125–60.

Klassen, Anne, Andrea L. Pusic, Amie Scott, Jennifer Klok, and Stefan J. Cano. 2009. "Satisfaction and Quality of Life in Women Who Undergo Breast Surgery: A Qualitative Study." *BMC Women's Health*, 9: 11.

McClimans, Leah. 2010. "A Theoretical Framework for Patient-Reported Outcome Measures." *Theoretical Medicine and Bioethics* 31: 225–40.

McClimans, Leah. 2011. "The Art of Asking Questions." *International Journal of Philosophical Studies* 19(4): 521–38.

McDowell, Ian. 2006. *Measuring Health.* 3rd Edition. Oxford: Oxford University Press.

Morgan, Mary, and Margaret Morrison, eds. 1999. *Models as Mediators*. Cambridge: Cambridge University Press.

Murray, D. W. et al. 2007. "The Use of the Oxford Hip and Knee Scores." *The Journal of Bone and Joint Surgery* 89B: 1010–14.

Rapkin, Bruce, and Carolyn Schwartz. 2004. "Toward a Theoretical Model of Quality-of-Life Appraisal: Implications of Findings from Studies of Response Shift." *Health and Quality of Life Outcomes* 2: 16.

Stenner, A. Jackson, and Donald S. Burdick. 1997. *The Objective Measurement of Reading Comprehension: In Response to Technical Questions Raised by the California Department of Education Technical Study Group*. Durham, NC: Metametrics, Inc.

Stenner, A. Jackson, William P. Fisher Jr., Mark H. Stone, and Donald S. Burdick. 2013. "Causal Rasch Models." *Frontiers in Psychology* 4: 1–14.

Stewart, Anita, and John Ware. 1992. *Measuring Functioning and Well-Being: The Medical Outcomes Study Approach*. Durham, NC: Duke University Press.

Streiner, David L., Geoffrey R. Norman, and John Cairney. 2015. *Health Measurement Scales: A Practical Guide to Their Development and Use.* 5th Edition. Oxford: Oxford University Press.

Tal, Eran. 2012. "The Epistemology of Measurement: A Model Based Account." (Doctoral dissertation). University of Toronto.

Van Fraassen, Bas. 1980. *Scientific Representation: Paradoxes of Perspective*. Oxford: Oxford University Press.

Westerman, Marjan J., Tony Hak, Mirjam A. G. Sprangers, and Anne-Mei The. 2008. "Listen to Their Answers! Response Behaviour in the Measurement of Physical and Role Functioning." *Quality of Life Research*, 17: 549–58.

Chapter 8

Measuring the Pure Patient Experience

Exploring the Theoretical Underpinnings of the Edmonton Symptom Assessment Scale

Eivind Engebretsen and Kristin Heggen

The policy and practice of welfare is increasingly concerned with placing the individual at the center of decision making and delivery services, which respond to individuals' needs. A range of systems and tools are developed to help monitor and measure service quality and adjust treatment and care to individuals' needs. In this chapter, we will discuss fundamental presuppositions underpinning the measurement practice in health care. We will illustrate this practice using an internationally recognized assessment instrument used in the treatment and care of cancer patients. The tool is called the Edmonton Symptom Assessment Scale (ESAS) and was introduced in 1991 as a method for regular assessment of symptom distress (Bruera et al. 1991).

The ESAS consists of a survey form designed to assist in the assessment of pain, tiredness, nausea, depression, anxiety, drowsiness, appetite, well-being, and shortness of breath. The severity of each symptom at the time of assessment is rated from 0 to 10 on a numerical scale; patients complete the assessment regularly (often daily). The scores provide a clinical profile of symptom severity and provide important information for choice of treatment and care of the patient. The ESAS is intended to "bring out the patient's subjective experience of their situation, and [is] important both to identify symptoms and to evaluate actions taken" (IS-1529 2007, 18). The ESAS is based on the idea that the patient is an expert on his or her own suffering and that his or her subjective experience is the gold standard for treatment. At the same time, the ESAS represents a standardized and objective way of assessing and measuring each patient's subjective experience and level of symptom distress. Thus, the ESAS is based on the double aim of strengthening both the subjective and objective aspects of assessment to offer patients the best quality of health care. The method differs significantly from a clinical encounter, which is an "in-between" space and a meeting ground for persons defined in relational and interdependent terms.

The aim of this chapter is twofold. First, we analyze the ESAS form and the clinical guidelines for the use of this form and discuss the paradoxical ambition of objectively measuring a patient's subjective experience. Second, we question two fundamental presuppositions underpinning this double ambition. We argue that ESAS is premised on a belief that there is a pure subjective experience that exists objectively outside and beyond language and interpretation. Our reflections and critique draw on Jacques Derrida's understanding of the relation between experience and language. From Derrida, we move to the Canadian philosopher and theologian Bernhard Lonergan as we discuss the second presupposition embedded in this form of measurement of symptom distress, namely, that an individual's knowledge about his or her experience is unique and private and can best be captured by reducing and controlling social interaction and interference.

ANALYSIS OF ESAS

The aim of ESAS is to "capture the patient's perspective on symptoms" (Alberta Health Services 2010, 1). Although, according to its guidelines, the ESAS is "only one part of a holistic clinical assessment" and "not a complete symptom assessment in itself" (Alberta Health Services 2010, 1), it relies on the assumption that pain experiences can be captured better through systematic measuring than in a less structured clinical encounter or dialogue. Studies have shown that patients are reluctant to report symptoms in clinical encounters but health workers tend to presuppose that patients will voluntarily report symptoms themselves (Paice 2004). Standardized assessment tools are supposed to overcome such barriers and ensure more accurate documentation of patient symptoms.

The underlying assumption is thus that standardization of subjective pain experiences ensures accuracy. This is also stated explicitly in the guidelines: "to assess the current symptom profile as accurately as possible" (Alberta Health Services 2010, 3). In line with this purpose, the guidelines also stress the importance of a "valid use" of the form. The assessment should, for instance, always be done on the ESAS numeric scale and not be inscribed directly into the ESAS graph so as not to risk bias by comparison of symptoms over time. The "attention to the graphed historical trend may affect the current scores" and thus undermine the correctness of the assessment (Alberta Health Services 2010, 3).

In other words, to be "valid" the assessment should be performed in a controlled manner. Subjective and pure patient experiences (i.e., untouched by any form of interference or interpretation) can be grasped only through a formal and standardized procedure. Hence, the form is designed based on the assumption that patients speak more accurately and adequately through a standardized voice, which allows predefined categories to articulate their symptom distress. This also implies that certain subjective experiences are excluded as irrelevant or as noise. The form allows for only subjective differences that can be graded on a standardized scale. Other differences (such as type or quality of pain) are ruled out. It is therefore assumed that subjective symptom experiences can be more accurately described through a standardized language (consisting of a one-dimensional numerical scale and fixed, predefined categories) than through ordinary language. A language that emphasizes differences in only degree of pain while excluding other experiences and differences is presumed to be more accurate or pure.

Furthermore, to assure "accuracy," the form should reduce the caregiver's interpretation as much as possible. Ideally, the assessment should be performed by the patient himself or herself. If the patient cannot participate in his or her own symptom assessment, the caregiver should assess the symptoms "as objectively as possible." To ensure objective assessment, a list of "objective indicators" is provided:

- Pain—grimacing, guarding against painful maneuvers
- Tiredness—increased amount of time spent resting
- Drowsiness—decreased level of alertness
- Nausea—retching or vomiting
- Appetite—quantity of food intake
- Shortness of breath—increased respiratory rate or effort that appears to be causing distress to the patient
- Depression—tearfulness, flat affect, withdrawal from social interactions, irritability, decreased concentration and/or memory, disturbed sleep pattern
- Anxiety—agitation, flushing, restlessness, sweating, increased heart rate (intermittent), shortness of breath
- Well-being—how the patient appears overall (Alberta Health Services 2010, 2)

The concept of objective indicators implies that the caregiver's subjective interpretation should be ruled out as much as possible. The caregiver's subjective interpretation is a potential source of bias and a threat against accuracy. This threat can be counterbalanced using objective indicators. Furthermore, it presupposes that symptom experiences have an objective existence independent of the patient's own individual interpretations and descriptions of his or her pain. The form assumes that there is a pure pain experience that exists prior to language and categorization and that can be adequately measured given the right assessment tool. Nausea, tiredness, anxiety, and depression are distinct and clearly distinguishable conditions that exist independently of the categories we use to describe them. Such conditions are intuitively recognizable and do not necessitate interpretation. Symptoms such as pain, anxiety, and nausea thus appear to be unambiguous phenomena. They change in degree but not in type. What type of pain the patient may feel and whether it is the same today as it was the day before or is perceived as a different pain entirely is not conveyed. Such information falls outside the form's registering gaze. The form does not solicit information about the causes of fear and anxiety. Only the degree to which these emotions are felt is made interesting and relevant. Thus, it is also said that all anxiety and sadness can be treated equally. It is the level of anxiety that guides the intervention, not its underlying causes or nature.

By insisting that the patient perform the assessment himself or herself, the form also construes pain experience as something fundamentally individual. Pain belongs to a private space to which only the affected individual has access. The whole idea of the form is to record subjective experience from within this private space without introducing confusing interpretations. This can best be achieved when the patient is presented with the form and answers it according to the standardized questions without interference from the caregiver. If necessary, any dialogue and interaction with the caregiver should be performed in a controlled and standardized manner by building on exact guidelines with objective indicators. Hence, we understand that any informal dialogue or interaction with the caregiver represents a possible source of "contamination" of the pure and private space of experience that the form gives access to.

Summing up, the ESAS form is based on two fundamental presuppositions:

1. There is a pure, subjective experience that exists objectively outside and beyond language and interpretation.
2. An individual's knowledge about his or her experience is unique and private and can best be captured by reducing and controlling social interaction and interference.

Following, we will challenge these two presuppositions by building on the work of Jacques Derrida and Bernhard Lonergan.

FIRST PRESUPPOSITION: THERE IS A PURE, SUBJECTIVE EXPERIENCE THAT EXISTS OBJECTIVELY OUTSIDE AND BEYOND LANGUAGE AND INTERPRETATION

In his famous critique of Husserl's phenomenology, Jacques Derrida insists that there is no such thing as a pure experience in which the experience of the individual is present to itself. For Derrida, experience is fundamentally interpretational and always already contaminated by language. Although Derrida agrees with Husserl that the world is given to us through our experiences, through our conscious acts, and that our consciousness is directed or intentional, he refutes one of the principal assumptions of Husserl's phenomenology: that the world of pure experience is fragile and cannot tolerate interference from outside.

For Husserl, language is a constant threat to intuition and true experience. As Derrida writes of him, "The expressive and logical purity of the *Bedeutung* that Husserl wants to grasp as the possibility of the *Logos* is gripped, that is contaminated—*in fact and always* (*allzeit verflochten ist*) insofar as the *Bedeutung* is gripped within a communicative discourse" (Derrida 2011, 17–18). In other words, experience cannot be communicated to others without running the risk of being misunderstood and transformed. However, Husserl tries to find his way out of this dilemma by introducing the distinction between indication and expression. While an indication only points to something, like a knot on a handkerchief reminding its owner to go to the store, an expression is identical with its meaning, like a note saying "remember to go to the store." Being only indicative, indications require interpretation, while expressions are self-sufficient. Husserl recognizes that the distinction is not clear cut, and he admits that most expressions contain the quality of indications (i.e., no matter how clear, an expression necessitates interpretation and therefore runs the risk of being misunderstood). It is only by thinking quietly to oneself, in a kind of inner monologue (a silent voice), that expressions can possibly be identical to experience. But once the experience has been communicated, it is affected by language. Derrida explains Husserl's concept of pure expression as follows:

> The subject does not have to pass outside of himself in order to be immediately affected by its activity of expression. My words are "alive" because they seem not to leave me, seem not to fall outside of me, outside of my breath, into a visible distance; they do not stop belonging to me, to be at my disposal, "without anything accessory." (Derrida 2011, 65)

Contrary to Husserl, who is not willing to give up the ambition of a pure expression, Derrida refutes the concept all together. For Derrida, this concept of a pure and "living" voice reflecting real and untouched experience is a myth. Derrida's claim is based on two arguments.

First Argument

No matter how internal, the voice is always contaminated by something external. Even the inner monologue is made possible by language, which is handed over to the individual by someone other, from the outside. Even the most intimate dialogue is, in a sense, an interpretation of other external utterances. An expression is therefore never untouched and identical to itself but rather always based on repetition and interpretation. Consciousness is haunted by something other than itself, by a nonpresence. Absolute self-presence is impossible because it is made possible by something external; it is conditioned by an absence, namely, language and repetition. As Derrida explains, "the possibility of re-petition in its most general form—that is the constitution of a trace in the most universal sense—is a possibility which not only must inhabit the pure actuality of the now but must constitute it through the very movement of difference it introduces" (Derrida 2011, 67). To clarify, the individual voice can make sense (even to the individual himself or herself) only by repeating and bearing trace of general code or language that is external to the individual. Even when we speak quietly to ourselves, meaning is made possible only through the repeatability of signs. This goes for even the most private of all words, the "I." For the "I" to make sense it must be different from the I that pronounces it; it is not unique but general and repeatable. The "I" makes sense only by pointing to other "I"s, by referring to a code. This difference between "I" and I is, according to Derrida, the precondition for signification.

One might argue that this is also the case for pain and symptom experiences. To categorize and make sense of symptoms and pain, which are singular, the individual must turn to conventional categories and classifications, which are not singular. The singular experience can be expressed only through language, which is not singular. Moreover, without this external interference, without the nonsingularity of common categories, experiences of pain and symptoms can never become singular. They would simply not make sense. To recognize an experience as a specific experience (as nausea or anxiety for instance) means to recognize it as belonging to something that is not unique, to a common category. This is true for even the most private monologue. In a sense, the idea of pure experience and expression is deconstructed by the ESAS form itself: on one hand, the form intends to capture pure experience prior to language and interpretation as an objectively existing experience, which belongs absolutely to the body of the individual. On the other hand, however, it states that this internal experience can be captured only through outer interference, through the introduction of an external standard and objective indicators.

Second Argument

Expressions of pure experience are also made impossible by the passing of time. The voice can never hear itself in absolute simultaneity, in the same moment that it speaks. To hear oneself is to hear oneself as another, as a different voice—in a kind of echo (Tellmann 2003). Self-presence implies delay and distortion and a constant need for self-interpretation. Not even in an inner monologue is expression self-evident. The I who speaks and the I who listens are separated by a delay in time. Meaning is always only indicative and never fully present. For example, before I state a simple utterance, such as "I am thinking about work," it has already become untrue. As I say it, I have started thinking about something else, namely, what I am thinking about. This delay that separates the I from itself and the expression from its meaning is what Derrida calls *différance* (with an a). The word *différance* is Derrida's linguistic construction, playing on both the noun—*différence*—and the present participle of the verb *différer*—*différant*—meaning "delaying". By incorporating the verb into the noun, Derrida highlights that différance is not only a structure but also a structuring force. Building on Saussure, who claims that language is a system of internal differences in which every signifier refers to a signified without any reference to the "real world," Derrida adds that the signifier and the signified are always separated by a delay. Saussure sees language as a synchronic system where the sign draws its meaning from its differential relations to other signs (e.g., "red" has meaning by being different from blue, green, etc.). Derrida insists that language also is a diachronic structure in which the meaning of the sign is dependent on its infinite uses and interpretations. Hence, the sign is not a fixed unity but rather is fundamentally unstable. Thus, meaning is never fully present in the sign (i.e., not self-evident, as in the Husserlean expression) but rather indicative and open to interpretation.

This contingency also goes for the categories of the ESAS form. Nausea, anxiety, tiredness, and depression do not refer to transcendental signifiers with stable meanings. They are constantly changing phenomena. By referring to symptoms and pain in terms of "stable" categories that can be graded equally from day to day, the form undermines the constantly changing nature of our experience. My anxiety today is not necessarily the same as it was yesterday—not only because it has changed in degree but also because the cause might be different or it simply feels different. This "gliding of signifiers" is indirectly referenced by the form insofar as it is what the form aims to escape. The form and the objective indicators it contains are needed only because the patient's own words are not stable.

This means that the ESAS's underlying ambition of creating a pure language through objective indicators is destined to fail. A pure language simply does not exist. No matter how simple the categories and precisely we try to define them through objective indicators, their meaning will never be stable. As in any other linguistic system, the categories in the form are open to an unlimited number of interpretations. Hence, a number on the scale will necessarily be a random point in an endless chain of interpretations and not the final answer to the question of what symptoms the patient is experiencing.

Where does this leave us when it comes to the possibility of pain assessment? If there is no pure experience prior to language and no objective language through which experience can be mediated, is pain experience impossible? To answer this question, we will explore the second presupposition underpinning the logic of the ESAS form by drawing on Bernard Lonergan's theories of subjectivity and knowing.

SECOND PRESUPPOSITION: THE INDIVIDUAL'S KNOWLEDGE ABOUT HIS OR HER EXPERIENCE IS UNIQUE AND PRIVATE AND CAN BEST BE CAPTURED BY REDUCING AND CONTROLLING SOCIAL INTERACTION AND INTERFERENCE

Bernard Lonergan has developed a theory of the subject that fundamentally questions the notion of knowledge that underpins modern philosophy (Lonergan 1992). He refutes the Cartesian concept of the subject as a res cogitans separate from a res extensa. Descartes claims that man can have an intuitive knowledge about his own mental acts without any reference to the world outside himself. Material substance, on the other hand, is knowable only through sensitive extroversion (Devina 2008, 32). As opposed to the Cartesian idea of an ordinary transcendental subject that immediately knows itself before it knows other objects, Lonergan claims that any knowledge—including self-knowledge—is mediated by meaning: "Knowing . . . is achieved when we break from the world of the infant, from the world that is reached, seen, heard, into the world as mediated by meaning" (Lonergan 1996, 199).

On this view, the problem with Descartes, and with modern philosophers of knowledge in general, is that they tend to neglect the multifarious character of knowing: "knowing . . . is a compound of three components: an experiential, an intellectual, and a judicial component" (Lonergan 1996, 198). There is no intuitive knowledge of the self, as Descartes claims. According to Lonergan, all knowledge comes into being through a process consisting of three separate, but complementary, cognitive operations: experiencing, understanding, and judging. Experiencing is the first, but still only one, phase of this process. Experiences do not in themselves qualify as knowledge because they do not make sense. For our experiences (either of ourselves or external objects) to be meaningful, they must be processed conceptually and judicially. We come to know our experiences by adding on to them, that is by enriching them through the incorporation of knowledge that we already possess. Lonergan compares this process to how we, when writing equations, put x on one side and everything we know about it on the other. We use our acquired knowledge to capture the unknown (Lonergan 1992, 61). To make sense of our experiences, we must test how they fit with already acquired concepts and categories. Lonergan refers to this operation as the "understanding phase," during which we question our experiences. Our understanding is fundamentally dialogical; it comes about through asking and answering questions (Lonergan 1992, 33–34). Through questioning we come to know what we do not know (the known unknown). At the same time, we assume that knowledge can be obtained. This way, the question bridges the gap between what we know and what we do not know.

Still, even after having gone through this process, we have only drafts of knowledge. The understanding phase is a creative process of interrogation and thus possibly endless. This process can be ended only through a decision, an active judgment. We need to weigh up different possible interpretations to reach a conclusion. "Judging," which is the name Lonergan gives to this phase, is an argumentative process through which a conclusion is made. This phase also involves processing and adding on to our experiences. We begin by confronting the preliminary understanding of our experiences (our knowledge drafts) with "heuristic structures." These structures are basic assumptions that guide our inquiry; they are also "canons," or general rules, that "govern the fruitful unfolding of the anticipations of intelligence" (Lonergan 1992, 93) and thereby determine the validity of our assumptions and acquired knowledge.

Terry Tekippe (2003) explains heuristic structure as "an almost automatic way of generating and utilizing clues" like "the algebraist's particular habit of solving problems by announcing 'Let X be the required number'" (Lonergan 1992, 60). Lonergan identifies two dominant and very general heuristic structures in the history of knowledge: (1) a classical heuristic structure going back to Aristotle, which is characterized by a search for certain, necessary causes and (2) a statistical heuristic structure, typical of modern science, where the focus is no longer causal necessity but rather frequency and probability. Both of these very general assumptions or styles of reasoning are governed by rules—or what Lonergan refers to as "canons"—concerning (1) selection (What counts as data?), (2) operations (How are insights accumulated?), (3) relevance (Which aspects of the data are of interest?), (4) parsimony (What kind of explanation may be legitimately added to the data?), (5) complete explanation (What counts as an adequate explanation?), and (6) statistical residue (What is *not* captured by the explanation?) (Lonergan 1992, 63).

Of importance in our context, this process of mediating our experiences is highly dependent on norms, rules, and assumptions that are not unique to the individual and are far from private. Experiencing can be performed only by the individual alone. The subsequent phases of knowing, understanding, and judging, involve a collectivity of meaning. To know our experiences (whether they concern ourselves or external objects), we must relate them to collective concepts and cultural norms through which they come to make sense. The act of questioning through which knowledge comes about is not an internal conversation with myself alone. It is a dialogue with a whole community of meaning. Meaning, and thereby true knowledge, is mediated only in the transitional space of relatedness between persons. It is not by locking myself up that I find myself and come to know my own experiences. On the contrary, the only way to self-knowledge is to live in the in-between space and participate in dialogue with others. Through dialogue with others I can test out how my experiences fit with common categories and acquire the norms and values against which my understanding can be judged.

Hence, Lonergan describes knowing as a subjective *and* an interactive process. More precisely, it is interactive in terms of being subjective. The subject is, according to Lonergan, a dialogical entity (Devina 2008). The self is not a substance but an assemblage of actions and interactions. It is a process of constant mediation through understanding and judgment. This process of mediation leans on categories, heuristics, and norms that are not defined by the individual alone but rather presuppose social interaction. Not only the subject, but also the concept of "reality," is for Lonergan, essentially mediated and interactional. Reality is not the res extensa of an "already out there now real." Instead, reality is meaning making; it is the "consequence of intelligent inquiry and critical reflection" (Lonergan 1992, 413), and this process is—as we have seen—fundamentally social. Thus, Lonergan challenges a Cartesian dualistic understanding of the subject as separate from the world. All beings—myself and others—are products of social interaction; hence, we are persons defined in relational and interdependent terms.

Applied to the case of ESAS, this interdependence means that knowing about our own pain must be understood as a social act. Like any other act of knowing, it is performed through interaction with a community of meaning. Knowledge about our pain does not capture pure experience. It captures how this experience makes sense by forming a concept of pain and leading to judgments. This process is highly interactional. It entails asking and answering questions, testing out categories and assumptions, and drawing on a wide community of meaning. Such knowledge cannot be created in a secluded space but rather necessitates active dialogue and involvement with others. I do not come to know my own pain by inward isolated reflection but rather by turning outward and engaging in dialogue with others. This requirement means that dialogue and interaction are not threats to self-knowledge but rather its precondition.

To a certain extent, the ESAS form admits the dialogical nature of pain knowledge by presupposing that such knowledge can be better grasped by asking and answering questions using the form of a questionnaire. The interference of the other (and his questioning) is thus part of the acquisition of pain knowledge. But the form also marginalizes this interference by presenting it as a voiceless/faceless questioning. The questioning is assumed to grasp the pain knowledge more accurately because it is anonymous. The form thus sticks to an objective-subjective dichotomy, which Lonergan argues we ought to transcend.

Through Lonergan we can create a new basis for knowledge of pain experience that differs from the one that is assumed not only by the measurement form but also from the nihilism that often follows the postmodern deconstruction of the subject. There is a way to self-knowledge and self-assessment, but it does not pass through objective measuring of neutral representation of pure individual experience. Instead, self-knowledge and self-assessment come about through conversation and dialogue. The dialogue with the caregiver is therefore not a threat against accuracy and purity of experience. Rather, it is through this dialogue that an individual's experience of himself or herself comes to make sense.

REFERENCES

Alberta Health Services. 2010. *Guidelines for Using the Revised Edmonton Symptom Assessment System (ESAS-r)*. Canada.

Bruera, Eduardo, Norma Kuehn, Melvin J. Miller, Pal Selmser, and K. Macmillan. 1991. "The Edmonton Symptom Assessment System (ESAS): A Simple Method for the Assessment of Palliative Care Patients." *Journal of Palliative Care* 7: 6–9.

Derrida, Jacques. 2011. *Voice and Phenomenon: Introduction to the Problem of the Sign in Husserl's Phenomenology*. Translated by Leonard Lawler. Evanston: Northwestern University Press.

Devina, Edgar A. 2008. *The Ground of the Normative Force of Discourse: A Lonerganian Reconstruction of Habermas's Communicative Rationality*. Ann Arbor: ProQuest.

IS-1529. 2007. Sosial- og helsedirektoratet. *Nasjonalt handlingsprogram med retningslinjer for palliasjon i kreftomsorg IS 1529 Nasjonale faglige retningslinjer*. Oslo: Sosial- og helsedirektoratet.

Lonergan, Bernard. 1992. *Insight: A Study of Human Understanding*. Toronto: University of Toronto Press.

Lonergan, Bernard. 1996. "The Analogy of Meaning." In *Philosophical and Theological Papers 1958–1964*. Toronto: University of Toronto Press.

Paice, Judith A. 2004. "Assessment of Symptom Clusters in People with Cancer." *Monographs-National Cancer Institute* 32: 98–102.

Tekippe, Terry. 2003. *Bernard Lonergan: An Introductory Guide to Insight*. New York: Paulist Press.

Tellmann, V. A. 2003. "Gjenlyden av en taus stemme. Stemmen og fenomenet i lys av fortellingen om Narcissus og Ekko" [The Sound of a Silent Voice: The Voice and the Phenomenon in the Light of the Story of Narcissus and Echo]. *Norsk filosofisk tidsskrift* [*Norwegian Philosophical Journal*]. 227–37.

Chapter 9

Measurement, Multiple Concurrent Chronic Conditions, and Complexity

Ross E. G. Upshur

Medical thinking evolved successfully by the identification of diseases and evolving clinical and health system measures to examine trends and associated health outcomes. However, the traditional metrics associated with disease-specific approaches are being challenged in the early twenty-first century.

The demographic transition occurring in high-income countries is well recognized. Life expectancy has increased in almost all high-income countries and is now also evident in many middle- and low-income countries. This transition has been referred to as the rectangularization of mortality. Associated with the increasing age of populations is an increase in chronic diseases. So far there has been no rectangularization of the morbidity curve. Rather, what is evident is that with aging comes multimorbidity (i.e., the occurrence of more than one chronic disease in the same person). As recent studies have shown, multimorbidity is the rule, not the exception, in populations. However, disease measurement and health system and service planning still largely operate on a model where single diseases predominate (e.g., cancer, diabetes, cardiovascular disease, arthritis, etc.), as if persons did not experience one or all these conditions simultaneously.

The issue of how social determinants, such as income and occupation, influence these morbidities has also gone largely unrecognized. But social factors have profound influences on singular conditions, are often highly associated with multimorbidity, and frequently result in the creation of complexity. Currently, there are no adequate measures at the micro-, meso-, and macro-level that characterize multimorbidity and complexity despite their being paramount challenges to health care delivery, health services delivery, and health policy.

This chapter will document the challenges associated with current measurement deficiencies in this area and propose an outline of a strategy to redress these deficiencies with the hope of having a set of measures more calibrated to an ontology of health in the era of multimorbidity and complexity.

DEMOGRAPHIC TRANSITION

It is now evident that there is a global demographic transition occurring. In all health systems, there is evidence of a growing number of older adults. Demographic trends are such that the traditional population pyramid is no longer the norm. In high-income countries, such as Canada, the fastest-growing segment of the population is the age group greater than eighty years old. In Canada, the number of people over the age of sixty-five exceeds that of those under fourteen (Statistics Canada 2015). The World Health Organization (WHO) estimates that soon the number of people over the age of sixty-five will exceed that of the under-five category for the first time in recorded history. It also estimates that the number of people older than sixty-five will exceed 1.5 billion by 2050 (WHO and US National Institute on Aging 2011).

This demographic transition is significant for health systems. It has been noted for some time that in high-income countries, the population mortality curve has become progressively rectangular. The average age of death is occurring at a progressively older age over the past half century. The reasons for this transition are complex and only partly explained by increased effectiveness of health care. However, this increase in longevity has not been associated with a reduction in morbidity. In fact, this demographic transition has been associated with a growth in individuals living with multiple concurrent chronic conditions (MCCC) (also known as multimorbidity, or pluripathology). For example, Denton and Spencer (2010) show that in Canada the number of people with no chronic conditions decreases over the life course and the co-occurrence of chronic diseases increases with age. This has been demonstrated in every health system globally (Barnett et al. 2012). MCCC is also associated with a wide range of additional issues relating to the social determinants of health (SDOH) and interactions with mental health conditions. These additional considerations constitute complex patient populations (Lawson et al. 2013; Schaink et al. 2012).

It is clear now that the co-occurrence of chronic diseases is the most prevalent form of chronic disease (Tinetti et al. 2012). But the idea that MCCC is the rule and not the exception for both population and individual health has not yet fully been integrated into health system planning and health professions' education. To my mind, MCCC thus represents the preeminent challenge of health system planning and health professions' education in the early twenty-first century. One key dimension of this challenge relates to measurement.

A common refrain at meetings held among clinicians and health policy leaders is that you cannot manage what you cannot measure. This refrain is stated in such a manner as to be axiomatic, uncontroversial, and universally believed. It is also likely to be false. My concern in this chapter, however, is not to demonstrate the falsity of this claim. While measurement may not be essential to management, it is assuredly important at all levels of the health care system: micro, meso, and macro. This said, mismeasurement is equally important and leads to mismanagement. I will argue for this thesis by discussing how investments in measurement at the clinic, hospital, and health system levels have failed to adapt to the changing nature of the populations served. This failure is most evidently the case with aging populations afflicted with MCCC and the growing population characterized as "complex." The measurement challenges posed by this population raise important questions about how best to proceed in a culture of measurement when there is no assurance that critical events are capable of being well measured.

CLASSIFICATION SYSTEMS

Disease classification systems are employed globally to provide data to inform health systems, measure performance, and track trends in population health. As taxonomies, they describe the ontology of disease, setting the criteria for what counts and what can be counted as a disease. These taxonomies thus set the fundamental basis for measurement.

Modern health care systems have evolved over the past fifty years based on focusing on diseases found in individual patients managed by health care professionals or through systematic surveys of persons living in the community. As such, almost all health measurement employs classification schemes to track events and outcomes in individuals that then aggregate up to inform policy and population health. Most of the classifications used are based on such taxonomies as the International Classification of Diseases (ICD); the International Classification of Functioning, Disability and Health (ICF); or in the case of mental health, the *Diagnostic and Statistical Manual of Mental Diseases* (*DSM*).

The ICD is the most important and commonly used disease classification system for health systems globally. According to the WHO Nomenclature Regulations, all member states are required to use the most current ICD to report morbidity and mortality statistics. According to the WHO, "ICD is the foundation for the identification of health trends and statistics globally, and the international standard for reporting diseases and health conditions. It is the diagnostic classification standard for all clinical and research purposes. ICD defines the universe of diseases, disorders, injuries and other related health conditions" (WHO 2016).

The ICD is organized by an orientation to diseases according to bodily systems that are linked to the primary cause. Hence, sections are devoted to infections, neoplasms, and diseases of body systems, such as the endocrine, central nervous, and cardiovascular system. This classification in turn reflects the organ-oriented approach taught to health professionals at schools of medicine and nursing as well as the organization of large health sciences centers, which have departments with highly specialized clinicians well trained to address diseases that arise in these discrete systems. The underlying assumption is that all conditions can be uniquely ascribed to the categories based on diagnostic criteria that permit sorting.

DOES THE ICD RECOGNIZE MCCC/COMPLEXITY?

As the foundation for the collection of diseases and health conditions that describes the "universe" of said entities, it would be important for this classification to capture what is now evidently the most prevalent form of chronic disease. However, the ICD is silent on the description of the co-occurrence of diseases. The ICD-10 does have one chapter devoted to "ill-defined" or "less well defined" entities:

> Signs and symptoms that point rather definitely to a given diagnosis have been assigned to a category in other chapters of the classification. In general, categories in this chapter include the less well-defined conditions and symptoms that, without the necessary study of the case to establish a final diagnosis, point perhaps equally to two or more diseases or to two or more systems of the body. Practically all categories in the chapter could be designated 'not otherwise specified', 'unknown etiology' or 'transient'. (ICD-10, Chapter XVIII)

It is further stipulated that the conditions and signs or symptoms included in categories R00–R99 consist of:

- Cases for which no more specific diagnosis can be made even after all the facts bearing on the case have been investigated.
- Signs or symptoms existing at the time of initial encounter that proved to be transient and whose causes could not be determined.
- Provisional diagnoses in a patient who failed to return for further investigation or care.
- Cases referred elsewhere for investigation or treatment before the diagnosis was made.

- Cases in which a more precise diagnosis was not available for any other reason.
- Certain symptoms, for which supplementary information is provided, that represent important problems in medical care.

Here, co-occurrence is a problem wherein it cannot be determined which single category is most responsible. That is, it is a problem of uncertainty. There is no sense in which the coexistence of two (and often more) conditions simultaneously constitute a legitimate category.

So as it stands, MCCC is nonexistent in that no such nomenclature that identifies it as an entity exists. Using the searchable index for the ICD does not bring up any relevant category to describe what increasing epidemiological research and expert opinion have labeled the most common and important health challenge in contemporary medicine.

WHY DOES THIS MATTER?

Clearly, if the most commonly employed classification system whose stated purpose is to describe the universe of diseases and health conditions that afflict humankind fails to acknowledge the existence of a major trend in health or does not capture the relevant event structures occurring in populations, the measures developed in conjunction with this system will be distorted in some important way. For instance, the measures will no longer be appropriately calibrated to the dynamics and structure of the health events in the populations they seek to understand. The short- and long-term consequences of this mismeasurement may be quite substantial. This lack of calibration may be even worse than the problems of the Gregorian calendar!

Clinicians and health systems currently rely on a wide range of data to characterize patient populations. These data can derive from many sources, such as electronic medical records, health records from health care institutions, and administrative data. Most of the measures used in health care are created from these sources, some at the point of care and some in summary form, such as discharge abstracts. For the most part, this "at-hand data" serve a multiplicity of purposes from financing through to research. Much of the data used in health care are collected for specific purposes in day-to-day care and are not yet well purposed to measure the sorts of events that are currently occurring in health care. As Cartwright and Bradburn (2011) observe:

> But purpose-built concepts make the accumulation of knowledge difficult so we are often reasonably pulled to rely on concepts that do not quite fit and to hope that the ones we use are close enough for results established in one situation to

> be reasonably accurate in others. It takes a serious case-by-case scientific decision to determine when this use can be expected to work well enough and when not and what if any the losses will be.

In the context of MCCC the problem to solve is to show how using measures that no longer quite fit, but are nonetheless being used on a regular basis to determine clinical and system performance, can be reengineered to better purpose.

THE PROBLEM OF OUTCOMES

Modern health systems widely proclaim allegiance to the triple aim: they seek to improve population health outcomes, increase patient satisfaction, and contain the cost curve. Many health systems are pledged to quality improvement initiatives, all devoted to improving outcomes. Indeed, many health systems are pledged to being outcome driven (Upshur, Kuluski, and Tracy 2014).

In the context of clinical care, outcomes are broadly considered to be the results brought about by the care delivered to patients by health care providers. At the population level, outcomes are understood in terms of metrics, such as rates of disease, mortality rates, and other measures of interest. In theory, at least, outcomes are closely related to the desired health-related goals of individuals and populations.

Understanding what exactly constitutes a "better" outcome presents a significant challenge that is relevant to achieving the triple aim. If improving population health outcomes is a central goal of the triple aim, then this improvement must link to the demographic transition in some systematic way. Moreover, these outcomes also claim to be person centered, meaning they are structured to address the needs of any given patient seeking care in the system. "Patient-centered" measures mean that the experiences (preferences and needs) of patients in this population must be systematically reflected in the measures that show how the population's health has been improved. There is also reason to believe that the SDOH, such as socioeconomic status, particularly deprivation and poverty, play a role in MCCC. People with less advantage tend to acquire MCCC younger and have more mental health conditions associated with their disease burden (Barnett et al. 2012). While the importance of these issues has been acknowledged, the measurement challenges have gone for the most part unaddressed.

Currently measured outcomes typically relate to the management of specific diseases, such as cancer, cardiovascular disease, and other common chronic diseases, such as diabetes, osteoarthritis, and depression. This is in keeping with the classification system set out in the ICD. But these outcomes neglect the experience of people with MCCC and fail to integrate SDOH. Poverty is not included in the ICD despite its evident relationship with increased risk and worse outcomes associated with many diseases (Commission on Social Determinants of Health 2008).

Disease-specific outcomes do not capture the experience or concerns of this population, who do not tend to place weight on any one of the chronic diseases they are afflicted with. Thus, in the context of MCCC these outcomes are not patient centered, and thus they miss the point of what tends to be most important to older patients and their families, namely, quality of life and optimized function and independence.

Older adults with MCCC suffer from a high level of symptom burden and complex care regimens involving multiple care providers. The research literature indicates that current health systems serve them poorly (Bierman and Tinetti 2016; Kuluski et al. 2016). For instance, care is not well coordinated between community providers and health care institutions, communication between providers is poor to absent, and patients and their family caregivers must navigate a confusing number of appointments and providers. Moreover, treatment goals or patient-centered outcomes related to a patient's needs are seldom acknowledged or discussed, and social factors are all but ignored.

There is a sparse literature on both measurement and outcomes in MCCC. Studies focusing on patient-relevant outcomes place less emphasis on managing diseases and a greater focus on optimizing function, reducing symptoms, and preserving independence particularly in the oldest old or among those with the highest morbidity burden. It is worth noting, however, that there is no unanimity regarding what specific outcomes are most appropriate in this patient population.

MCCC NOT THE ONLY ISSUE; COMPLEXITY NOW A FACTOR

MCCC is but one component of patient complexity. Schaink et al. (2012) developed a framework for understanding additional factors that contribute to complexity. They include patterns of health care utilization (e.g., high utilization), psychosocial factors (e.g., mental health addictions), and impairments of cognition and challenges in the SDOH (e.g., income, housing, and food security). Complex patients often have several interacting factors at play that thwart simple solutions based in medical approaches.

A complexity framework provides not only guidance in terms of explaining the nature of complexity but should also be directive in helping to conceptualize the variety of care needs that should be anticipated when seeking to provide care. Unfortunately, most clinicians and other service providers come equipped with skills that will only partially address these many needs and are often at a loss when it comes to figuring out the most appropriate way to address multiple needs. Current systems of measurement fail to integrate consideration of complexity as it relates to populations.

CHALLENGES IN MEASUREMENT

An adequate measurement system that can accommodate MCCC and complexity would require a significant reconceptualization of the things that are measured by clinicians, health planners, and policy makers. We may in fact need a new universe to be described. Given the dramatic changes in the population, the increasing recognition of the limitations of certain definitions of disease, and the importance of social determinants in shaping and modifying health, the ICD classification increasingly resembles a Ptolemaic universe in a Copernican environment. This is particularly the case with increasing calls for health care to embrace patient centeredness and patient-oriented outcomes. This moves the measurement agenda away from purely biomedical classifications to engaging with patient perspectives and their attendant social relationships.

This move toward a new universe, rotating around the patient, while appealing and desirable in many ways, poses significant challenges if concerns for measurement theory are not taken seriously. This move will entail commitment on two fronts. First, we need to have concerted focus on the desiderata of measurement. This focus will require explicit recognition of the epistemological issues involved in measurement and will necessitate attention to conceptual thinking and not simply endorsing available data sources or current measurement theory as adequate to the task. A second requirement is that leaders in health systems must invest in efforts to ensure that measurement issues are addressed thoughtfully and resources are available to develop, implement, and evaluate new measures. This should not be the exclusive domain of academia and the research world. Patients, clinicians, managers, and policy makers will continue, as they do now, to use the same data to inform decisions and allocate resources.

One of the virtues of many of the entry lines of the ICD is their clear definition and precision. Therefore, my proposal is not to abandon the ICD classification entirely but rather to recognize better its limitations and enhance it with a taxonomic approach that more accurately reflects the events occurring within health systems (Applegate and Ip 2016). We would do well to take note of cautionary advice from Nancy Cartwright and Norman Bradburn (2011, 18):

> Many policy-related social science concepts are value-laden or at least exactly how they are defined and measured will have many value-relevant consequences. Often we lack a firm scientific basis for important choices in how they should be defined and measured. The varying purposes for which they are used and the varying values assigned to the consequences make common metrics very difficult in these cases, if not impossible.

Many of the properties that need to be measured in MCCC/Complexity patient populations are value laden and have many value-relevant considerations. This becomes increasingly the case the more one moves away from the purely biomedical and into the social/functional domains. An additional complication is the need to have measures that integrate divergent perspectives on the same issue. I will use the example of goal setting to illustrate the complexities associated with generating meaningful measures that can inform management of complex patients.

In a qualitative study, Kuluski and colleagues (2013) used depth interviews to understand the extent to which patients and their caregivers and primary care clinicians shared the same goals of care. Goal setting is an important component of determining appropriate care particularly for older adults with significant burden of chronic disease.

The vast proportion of health care for complex older adults occurs in the primary care setting. Primary care is well situated to address the unique needs of each patient to provide patients and their family caregivers with the tools to manage their illnesses. It has become evident that primary care struggles to manage complex older adults with multimorbidity. Clinical practice guidelines that are primarily designed for single diseases have limited applicability for persons with multimorbidity (Mutasingwa, Ge, and Upshur 2011). Physicians may be forced to make decisions that involve prioritization or trade-offs, warranting discussions with patients on what is important to them and what they would like to achieve in terms of their health (i.e., goal setting). Understanding patients' goals of care can potentially aid in the successful management of their diseases at home and when integrated into care plans, can improve their quality of life.

Goal setting is not necessarily a formal part of primary care practice. For example, a Canadian survey of the experiences of primary health care patients drawn from a nationally representative sample of 11,582 patients with at least one chronic condition found that less than half of patients (48 percent) talked to their health care provider (at least some of the time) about their treatment goals (Canadian Institute of Health Information 2011). International data involving goal setting between physicians and adults with chronic illness have noted similar trends (Schoen et al. 2009). Although rarely studied, some research has started to elucidate what patient goals may look like.

Failure to share goals raises a risk that physicians may focus on aspects of care and treatments that are not desired by the patient and/or family member. Conversely, the patient and family may focus on things that the physician does not deem feasible. In interviewing twenty-seven triads consisting of patient, caregiver, and primary care clinician, the researchers found that at the aggregate level, common goals were expressed. The goals were broadly patient centered: maintenance of functional independence of patients and the management of their symptoms or functional challenges. However, despite common goals at the aggregate level, little alignment of goals was found when looking across patient-caregiver and physician triads. Lack of alignment tended to occur when patients had unstable or declining functional or cognitive health, safety threats were noted, and enhanced care services were required.

What this study suggests is that aggregate-level metrics looking at goal articulation will miss the fact that goals may be misaligned among those needed to achieve the goal. A patient-centered metric that is uniperspectival will miss the differential interpretation of the same measure by people involved in the care of the same patient. In some ways, the idea of goal articulation bears the features of a quantity that can be well characterized and measured. Indeed, it is amenable to a variety of potential metrics from dichotomous (goals articulated yes/no) to Likert-type scales.

But there is no straightforward way to capture goal misalignment. Thus, arguments that state that functional accounts for managing patients should be the dominant strategy will need to take into consideration the multiperspectival nature of the phenomenon of goal setting. This ambition leads to an important question: Is the purpose of the metric consistent among stakeholders and users?

Clearly, it is important to have measurement of patient-related phenomenon built from the ground up. However, patients do not exist in isolation from their health care providers and caregivers. The experiences of caregivers and providers are equally important; thus, measurement efforts that condition on patient perspectives alone will only partially represent the phenomenon. The problem becomes even more daunting when one factors in the reality that complex older adults often see a wide range of health care providers: specialist physicians; nurses; pharmacists and allied health care professionals, such as physiotherapists, dieticians, and social workers; and others. Furthermore, to be maximally useful, measures should additionally capture the influence of SDOH in this population.

The complex web of relationships is often greatly underestimated, and measurement efforts that fail to attempt representation of this complex matrix of relationships, communication, and information exchange will likely distort the phenomenon. It is critically important that measurement efforts honestly grapple with these challenges rather than defaulting back to what can be measured with extant data sources.

THE BALLUNG NATURE OF MCCC/COMPLEXITY

Cartwright and Bradburn (2011) propose a three-step approach to measurement:

1. Characterization: What concept is being measured?
2. Representation: define the properties of the measurement scheme.
3. Procedures: formulate rules for applying the scheme.

MCCC/Complexity poses significant challenges for the first of these (i.e., characterization). Currently, there is no agreement on what counts as MCCC. Different research groups use different thresholds of chronic conditions. Simple counts are used, such as two, three, or four or more chronic conditions. What qualifies as a chronic condition varies as well. Some groups include only identified ICD categories of chronic diseases, such as hypertension, diabetes, and coronary artery disease. This approach neglects symptom complexes, such as fatigue, insomnia, and malaise, which are common in these populations, chronic in nature, and relevant to patient concerns. Furthermore, adding in the complexities associated with SDOH makes straightforward measurement difficult.

A critical challenge for measurement identified by Cartwright and Bradburn (2011, 4) is the potential ambiguity of many concepts in health care. Drawing from Otto Neurath, they refer to this as Ballung concepts:

> Otto Neurath (socialist, sociologist, one of the founding members of the Vienna Circle, and spearhead of the unity of science movement of the 1930's) maintained that most concepts used in daily life are of the second type. He called them *Ballungen* ('congestions'), as in the German "Ballungsgebiet" for a congested urban area with ill-defined edges: There is a lot packed into it; there is often no central core without which one doesn't merit the label; different clusterings of features from the congestion (*Ballung*) can matter for different uses; whether a feature counts as in or outside the concept, and how far, is context and use dependent. We employ Neurath's word here because other words more commonly in use, like 'umbrella concept', generally already have interpretations, and different ones for different scholars and different fields.

MCCC and complexity are most assuredly congested, value-laden, and fuzzy concepts. But the critical question is whether they can be used in a meaningful sense to describe individuals and populations. As has been argued thus far, there is an undeniable sense in which MCCC/Complexity denotes a phenomenon occurring in health systems. There is also considerable debate as to the precise meaning of such concepts in terms of what is included in the scope and ambit of the application.

As noted above with the example of goal identification and alignment, there is an inherent information complexity and interpretive element to capturing many common events in the life world of patients afflicted with MCCC. With widely varying ideas on the basic conceptualization of the phenomenon, it may not be possible to contemplate yet the remaining steps of representing the concepts or developing procedures to apply to the measured property.

It will be no simple task to disentangle or decongest the concept of MCCC/Complexity. Despite an urgent need for measures that capture the phenomenon, we may be at a stage where we cannot in any meaningful way measure it. Currently, we may be in a particularly difficult position where we need to overcome neglect of the phenomenon yet are in no position to put in place a reliable measurement system to capture it. The perils of this difficulty are evident, and thus in conclusion, I will make a plea for urgent attention to this situation.

MOVING FORWARD: A MEASUREMENT AGENDA

The first step toward better measurement is awareness that current systems are no longer adequate. This situation is now receiving attention. Many scholars have noted the limitations of single disease approaches in the clinical domain and criticized efforts to base performance metrics for the MCCC/Complexity patient population on such measures. I will focus on the work of Mary Tinetti, as she has been an influential thought leader in sensitizing health care to the limits of current approaches, both clinically and regarding measurement for MCCC/Complexity populations. For example, Tinetti, Fried, and Boyd (2012, 2493) have stated:

> Initiatives by the Centers for Medicare & Medicaid Services (CMS) and private insurers designed to pay clinicians for quality, not merely quantity, of services hold promise for individuals with multiple chronic conditions. However, the initial CMS hospital-based metrics foster adherence to disease-specific (e.g., myocardial infarction, community-acquired pneumonia, heart failure) or procedure-specific (e.g., surgery) processes. These metrics encourage continuation of fragmented disease-centric care. None of the measures specifically address issues faced by patients with multiple chronic conditions.

This quotation clearly indicates the need to question measures and refocus attention to the life world of patients with MCCC. It fails to recognize some of the inherent issues implicit in measurement of these issues. It is one thing to recognize the limitations of the current measurement scheme but quite another to face the implicit and perhaps insurmountable challenges that may be faced.

Tinetti, Esterson, et al. (2016, 9) recently reported an initiative to address the current shortcomings of current systems of care. They convened an advisory group of "patients, caregivers, clinicians, health system engineers, health care system leaders, [and] payers" and

> identified three modifiable contributors to this fragmented, burdensome care: decision making and care focused on diseases, not patients; inadequate delineation of roles and responsibilities and accountability among clinicians; and lack of attention to what matters to patients and caregivers (i.e., their health outcome goals and care preferences). The advisory group identified patient priority-directed care as a feasible, sustainable approach to addressing these modifiable factors.

Tinetti and colleagues have identified an important step of moving forward to a new way of constructing measurement systems for MCCC/Complexity. Convening a diverse advisory group is a necessary first step to identifying the elements. Although some consensus was reached on what is feasible, it is critically important that such efforts place specific focus on measurement at the individual, system, and population levels. The idea of moving to patient priority–directed care is a promising one, and its success or failure will depend on the degree to which it can commensurate the various stakeholder perspectives and come up with a measurement system that addresses the Ballung nature of the very idea of patient priority–directed care itself.

The next step would be to create a taxonomy like the ICD system that reflects this move away from disease-specific approaches. A suggested template has been created by Tinetti, Naik, and Dodson (2016). They identify domains of health goals that relate to the following domains experienced by patients with MCCC:

- Function (e.g., walk two blocks without shortness of breath; live in my own home until I need help from someone at night).
- Symptoms (e.g., reduce back pain enough to perform morning activities without medications that cause drowsiness; get my appetite back and be able to eat the foods I like).
- Life prolongation (e.g., see my grandson graduate from high school in five years).
- Well-being (e.g., be as free from anxiety or uncertainty about cancer recurrence as possible).

- Occupational/social roles (e.g., work three more years; pick up my granddaughter from school).

These goals work at the individual level but fail to include the framing SDOH that clearly interact with and enable individuals to realize their health goals. They presume access to care and stable environments, such as housing and food security, which are critical dimensions. Thus, they are necessary but not sufficient to characterize the measurement schemes required to adequately capture this patient population.

If the concepts entailed in the characterization of MCCC/Complexity are inherently and inexplicably *Ballungen*, then measurement efforts will need to take this into account from the outset. Currently, there is little recognition of the need to create new measures for MCCC/Complexity, let alone acknowledgment of the challenges that the *Ballungen* nature of the phenomena associated with MCCC/Complexity poses. This lack of acknowledgment in and of itself does not argue against measurement in this field but rather should direct attention to efforts focused on closer attention to definitions. In this nascent field, the next steps will likely be dominated by Delphi exercises, consensus conferences, and expert panels. They should all be charged with paying close attention to conceptualization efforts so that the next steps of representation and processes of making the mathematics operational can take place. There is reason to be skeptical that concepts can be defined to permit the sort of precision demanded by measurement theory. It may be appropriate to open the discussion to how much precision is required or whether a certain degree of fuzziness can be tolerated. Clearly, despite the urgent need for new metrics, much work is required. Much of this work is philosophical in nature and if done properly, will reorient measurement to appropriately reflect the phenomenon it seeks to capture.

CONCLUSION

I have argued that there is considerable work required to reform measurement practices to better reflect the demographic and epidemiologic transition currently occurring globally. The ICD universe needs to expand, as it is currently blind to a major health trend. Proposed initiatives to focus primarily on patient-based accounts will falter, as they will not consider the equally important perspectives of clinicians and caregivers. This is particularly important as MCCC burden increases or cognition-impairing conditions, such as dementia, predominate. SDOH, which are a major factor in the occurrence of conditions and in outcomes, need to be systematically included for both scientific and justice considerations.

Resource allocation, human health resource planning, surveillance, and payment schemes are based on classification and measurement schemes. As I have argued, they are no longer adequately calibrated to the phenomenon occurring at the patient and population level. An additional concern for future measurement efforts are the challenges posed by the Ballung nature of the things being measured. There will need to be a multiplicity of measures developed, but more importantly, there may need to be a considerable discursive space to navigate rival uses and interpretations.

REFERENCES

Applegate, William B., and Edward Ip. 2016. "The Evolving Taxonomy of Health in Older Persons." *Journal of the American Medical Association* 316: 2487–88.

Barnett, Karen, Stewart W. Mercer, Michael Norbury, Graham Watt, Sally Wyke, and Bruce Guthrie. 2012. "Epidemiology of Multimorbidity and Implications for Health Care, Research, and Medical Education: A Cross-Sectional Study." *Lancet* 380: 37–43.

Bierman, Arlene S., and Mary E. Tinetti. 2016. "Precision Medicine to Precision Care: Managing Multimorbidity." *Lancet* 388: 2721–23.

Canadian Institute for Health Information. 2011. *Seniors and the Health Care System: What Is the Impact of Multiple Chronic Conditions*. Analysis in Brief. Ottawa, Ontario: Canadian Institute for Health Information.

Cartwright, Nancy, and Norman Bradburn. 2011. "A Theory of Measurement." https://www.researchgate.net/publication/283420488_A_Theory_of_Measurement. Accessed June 4, 2017.

Commission on Social Determinants of Health. 2008. *Closing the Gap in a Generation: Health Equity through Action on the Social Determinants of Health*. Final Report of the Commission on Social Determinants of Health. Geneva: World Health Organization.

Denton, Frank T., and Byron G. Spencer. 2010. "Chronic Health Conditions: Changing Prevalence in an Aging Population and Some Implications for the Delivery of Healthcare Services." *Canadian Journal on Aging* 29: 11–21.

Kuluski, Kerry, Ashlinder Gill, Gayathri Naganathan, Ross Upshur, R. Liisa Jaakkimainen, and Walter P. Wodchis. 2013. "A Qualitative Descriptive Study on the Alignment of Care Goals between Older Persons with Multi-Morbidities, Their Family Physicians and Informal Caregivers." *BMC Family Practice* 14: 133.

Kuluski, Kerry, Allie Peckham, A. Paul Williams, and Ross Upshur. 2016. "What Gets in the Way of Person-Centred Care for People with Multimorbidity? Lessons from Ontario, Canada." *Healthcare Quarterly* 19: 17–23.

Lawson, Kenny D., Stewart W. Mercer, Sally Wyke, Eleanor Grieve, Bruce Guthrie, Graham C. Watt, and Elizabeth A. E. Fenwick. 2013. "Double Trouble: The Impact of Multimorbidity and Deprivation on Preference-Weighted Health Related Quality of Life: A Cross Sectional Analysis of the Scottish Health Survey." *International Journal for Equity in Health* 12: 67.

Mutasingwa Donatus, Hong Ge, and Ross E. G. Upshur. 2011. "How Applicable Are Clinical Practice Guidelines to Elderly with Co-Morbidities?" *Canadian Family Physician* 57: e253–62.

Schaink, Alexis, Kerry Kuluski, Renée F. Lyons, Martin Fortin, Alejandro R. Jadad, Ross Upshur, and Walter P. Wodchis. 2012. "A Scoping Review and Thematic Classification of Patient Complexity: Offering a Unifying Framework." *Journal of Comorbidity* 2: 1–9.

Schoen, Cathy, Robin Osborn, Sabrina K. H. How, Michelle M. Doty, and Jordan Peugh. 2009. "In Chronic Condition: Experiences of Patients with Complex Health Care Needs, in Eight Countries." *Health Affairs* 28: w1–w16.

Statistics Canada. 2015. "Canada's Population Estimates: Age and Sex, July 1, 2015." Last modified September 29, 2015. http://www.statcan.gc.ca/daily-quotidien/150929/dq150929b-eng.htm.

Tinetti, Mary E., Jessica Esterson, Rosie Ferris, Phillip Posner, and Caroline S. Blaum. 2016. "Patient Priority-Directed Decision Making and Care for Older Adults with Multiple Chronic Conditions." *Clinics in Geriatric Medicine* 32: 261–75.

Tinetti, Mary E., Terri R. Fried, and Cynthia M. Boyd. 2012. "Designing Health Care for the Most Common Chronic Condition—Multimorbidity." *Journal of the American Medical Association* 307: 2493–94.

Tinetti, Mary E., Aanand D. Naik, and John A. Dodson. 2016. "Moving from Disease-Centered to Patient Goals-Directed Care for Patients with Multiple Chronic Conditions: Patient Value-Based Care." *Journal of the American Medical Association Cardiology* 1: 9–10. doi:10.1001/jamacardio.2015.0248.

Upshur, Ross, Kerry Kuluski, and Shawn Tracy. 2014. "Rethinking Health Outcomes in the Era of Multiple Concurrent Chronic Conditions." *Healthy Debates*. http://healthydebate.ca/opinions/rethinking-health-outcomes-in-the-era-of-multiple-concurrent-chronic-conditions.

World Health Organization. 2016. International Statistical Classification of Diseases and Related Health Problems. 10th Revision. Geneva: WHO Press. http://apps.who.int/classifications/icd10/browse/Content/statichtml/ICD10Volume2_en_2016.pdf.

World Health Organization and US National Institute on Aging. 2011. *Global Health and Aging*. Geneva: WHO Press. http://www.who.int/ageing/publications/global_health.pdf.

Part III

MEASUREMENT AND POLICY

Chapter 10

NICE's Cost-Effectiveness Threshold

How We Learned to Stop Worrying and (Almost) Love the £20,000–£30,000/QALY Figure

Gabriele Badano, Stephen John, and Trenholme Junghans

The National Institute for Health and Care Excellence (NICE) is a public body working at arm's length from the Department of Health in England. It is well known for its health technology appraisal (HTA) process, through which it assesses whether new drugs and other health technologies should be used in the National Health Service (NHS). If NICE recommends the use of a certain health technology, then clinical commissioning groups (i.e., local authorities responsible for allocating the NHS budget) are legally bound to fund it. These decisions can be extremely controversial. For example, in November 2015, NICE advised clinical commissioning groups against purchasing Kadcyla, a drug for late-stage terminal breast cancer sufferers. Although Kadcyla offers as much as an extra six months of life, it is highly costly, at around £90,000 for a fourteen-month treatment. This hefty price tag places Kadcyla well beyond the maximum amount of money that according to NICE, the NHS should pay per life year saved (NICE 2015). Many stakeholders expressed anger at this decision, with a representative of the charity Breast Cancer demanding a change to funding arrangements because "people living with incurable cancer don't have time to lose" (Boseley 2015).

What is the process behind this and many other controversial decisions by NICE? To decide whether to recommend a health technology, NICE collects evidence regarding its clinical effectiveness. The Quality-Adjusted Life Year (QALY), which integrates gains in life expectancy with improvements in quality of life, is NICE's measure of choice for determining the health benefits that a course of intervention can provide. The evidence about clinical effectiveness is brought together with evidence about financial costs to calculate the incremental cost-effectiveness ratio (ICER) of the technology (i.e., the additional cost of an additional QALY that the NHS would gain by using such technology compared to the health technology that the NHS currently employs for the same purpose). As a simplified example, imagine that a course of the current standard treatment for a certain form of terminal cancer costs £10,000 and its only effect is that, on average, it extends life by six months, without any improvement in quality of life. A new intervention could replace it, offering an average of twelve months of life extension to each patient at the cost of £15,000 per course of treatment. In this case, the ICER of the new treatment would be £5,000 (£15,000 minus £10,000) per six (twelve minus six) quality-adjusted months, or £10,000/QALY.

A key step in NICE's decision making process is the comparison between the ICER of the technology under appraisal with a cost-effectiveness threshold of £20,000–£30,000/QALY. Indeed, NICE explains that it is unlikely to reject any technology whose ICER lies below £20,000/QALY. If the ICER falls above £20,000/QALY, NICE's committees must reach beyond cost effectiveness and consider factors including the so-called "equity weightings" (severity of target disease, the innovative nature of the technology, extra priority to be assigned to end-of-life care, a premium to be placed on the treatment of diseases that disproportionately affect children or members of disadvantaged social groups). If the ICER of the technology under appraisal is between £20,000 and £30,000 per QALY, some of these factors must lend support to its use for such technology to be approved by NICE. Above the £30,000/QALY mark, an exceptionally strong case must be built on such factors if decision makers wish to recommend the technology despite its high ICER.[1]

Although NICE is not at liberty to disclose the exact figure, Kadcyla was rejected because of the large gap between its ICER, which falls in the region of £160,000/QALY, and the upper end of NICE's £20,000–£30,000/QALY threshold (NICE 2015). The public anger caused by NICE's use of its threshold to reject drugs is familiar. Generally, economists and ethicists also think that this anger is misguided, at least in one important sense. They argue that given that the resources devoted to health care are finite, the medical needs of our societies are virtually endless, and there is an extremely broad range of beneficial interventions available, there must be some beneficial drugs that health care providers will not purchase. Moreover, no decision to fund a new drug can simply be based on its clinical effectiveness, or in other words, on the fact that it would do good to patients. The clinical benefits that funding that drug would provide to patients must be compared with its "opportunity cost" (i.e., the clinical benefits to other patients that would have to be forgone by diverting the necessary funds from somewhere else in the NHS). One obvious way in which to do this is by comparing treatments in terms of how many QALYs they would generate for the money spent on them and then allocating funds based on a strong concern for cost effectiveness.[2] None of this is to say, of course, that the specific £20,000–£30,000/QALY threshold used by NICE—and, hence, the Kadcyla decision—is correct, nor is it to say that cost effectiveness is the only relevant consideration in this, or other cases of, just resource allocation. It is, however, to say that there must be something like a cost-effectiveness threshold beyond which the purchase of a drug will become increasingly unlikely even if this means that those without time to lose avoidably lose that time. In other words, the consensus among economists and ethicists is that one possible critique of NICE's £20,000–£30,000/QALY threshold—that it aims to measure something that should not concern resource allocation agencies in the first place—is off the table.

In this paper, we study the history of NICE's threshold as a way of investigating the additional steps necessary for a complete critical assessment of this threshold. We do this from the perspective of both ethics and philosophy of science. More specifically, our aim is twofold. In the first two sections (respectively, "The Threshold: The Theory" and "The Threshold: The Practice"), we engage in a close study of the theory and history of NICE's work. We argue that viewed as an attempt to measure its stated construct, the "opportunity cost" associated with funding a treatment, NICE's threshold—one of the most important measures in British public life, which quite literally determines life-and-death decisions—is deeply flawed. Therefore, NICE's £/QALY threshold badly measures its stated concern with cost effectiveness. But in the third section ("Justifying the Threshold?"), we argue that close attention to the complex institutional and political context of NICE's work suggests a more nuanced understanding of the role of this threshold, not only as a measure of the independent quantity that it explicitly sets to target, but also as a standard that serves to promote other important goods.

It is commonplace that the adequacy of some measures is related to moral and political ends and therefore that we cannot properly assess these measures without engaging in ethical debates. For example, one could challenge NICE's threshold by questioning the value judgment that NICE should be concerned with cost effectiveness at all. However, reflection on themes in philosophy, sociology, and anthropology of measurement allows us to consider the broader political context of these measures and the roles they might play beyond their stated ends. In turn, we can ask distinctively evaluative questions about those roles. This perspective points toward a way in which our assessment of measures in domains such as health policy must be sensitive to moral and political, as well as epistemological, questions. In this paper, we do not aim to say the last word about whether NICE's £20,000–£30,000/QALY threshold is justified. However, we certainly wish to highlight the sheer complexity of assessing this question and the ways in which even apparently flawed measures may do important political work.

THE THRESHOLD: THE THEORY

In philosophy, as well as in other disciplines, much attention has been paid to QALYs as a measure of health benefits.[3] The focus of this chapter is different in that we are interested in NICE's cost-effectiveness threshold quite independently of the choice of QALYs as its health outcome measure. Specifically, we are interested in the threshold's history and how it can be evaluated. To set the stage for our evaluation, this section explores the theory behind NICE's use of a cost-effectiveness threshold and the way in which theorists normally debate these thresholds.

In principle, NICE's £/QALY threshold is a measure of opportunity costs. However, before exploring this understanding, it is worth mentioning that an alternative view exists. According to the social-value view, NICE's threshold should measure the value that the British society at large attaches to one QALY's worth of health gain—a value that is given by the amount of money that the members of society are willing to pay to produce one extra QALY. Although commentators have sometimes used this approach to explain the meaning of NICE's threshold (Smith and Richardson 2005), this explanation suffers from serious flaws. An integral part of the social-value view is that the NHS should pay for all and only those interventions that produce a QALY for a cost that is equal or inferior to what has been found to be society's willingness to pay for it. This proposal suggests that NICE effectively determines the level at which the overall NHS budget should be set simply by setting its threshold. More realistically, however, the size of the health care budget is understood as an issue for parliamentary debate, to be settled based on a richer set of considerations than willingness to pay (McCabe, Claxton, and Culyer 2008, 735–36).

Indeed, NICE itself acknowledges that setting the overall level of public spending for health care is not its job and formally endorses an opportunity-costs understanding of its cost-effectiveness threshold. For example, in the latest edition of its *Guide to the Methods of Technology Appraisal*, NICE claims that a health technology is to be considered cost effective "if its health benefits are greater than *the opportunity costs* of programmes displaced to fund the new technology, in the context of a fixed NHS budget" (NICE 2013, 14; italics added). This view of cost effectiveness is grounded in the acknowledgment that available NHS resources are limited. Every recommendation that NICE makes in favor of a new technology that is costlier than the one adopted so far for the same use requires disinvestment somewhere else in the NHS. In this context, it is not enough that the new technology offers greater health benefits than the currently funded alternative; it is also important that the extra health benefits that would be obtained by commissioning it outweigh its "opportunity costs" (i.e., the health benefits that would be lost through disinvestment across the NHS). Very roughly, if a new treatment would cost an extra £40,000 to produce an extra QALY, this £40,000 cannot be used for other treatments, which might well produce more than one QALY (albeit, obviously, for different people). Consequently, it is important that the ICER of new technologies falls below the cost of a QALY produced through the least cost-effective intervention currently funded by the NHS. The £/QALY threshold is supposed to measure this cost (McCabe et al. 2008, 737–78).[4]

An interesting implication of this view of cost-effectiveness thresholds is that such thresholds should be frequently updated. In principle, if less cost-effective technologies are displaced over time by more cost-effective ones, NICE's £/QALY threshold should move downward and become more restrictive. Also, any real-term increase (or decrease) in the size of the health care budget should lead to a higher (or a lower) threshold.

The theory behind NICE's cost-effectiveness threshold is premised on the idea that a key goal of health care resource allocation decision makers is to use available funds to improve the aggregate health of the population, measured in terms of QALYs, as much as possible. In the philosophical debate over the ethics of health care resource allocation, there is broad consensus on the importance of this goal and, therefore, on the importance of measuring the opportunity costs of interventions in terms of displaced health benefits. However, there is also broad consensus on the idea that it is legitimate to pursue the goal of QALY maximization across society only under certain constraints, which express the importance of *who* receives the QALYs that are produced. For example, it is often argued that there are cases in which NICE and other health care resource allocation agencies should decide in favor of an intervention that they know will displace more QALYs than it will generate because this intervention will benefit patients who are particularly badly off (e.g., in terms of severity of illness, socioeconomic status, age).[5] These quintessentially distributive concerns are reflected in the equity weightings that we described in the introduction and that are balanced by NICE against the cost effectiveness (or lack thereof) of the technologies under appraisal.

THE THRESHOLD: THE PRACTICE

As explained in the previous section, a £/QALY threshold is meant to measure something rather specific—the cost of a QALY produced through the least cost-effective intervention currently funded by the NHS. We also saw that at least formally, NICE accepts this idea of what the cost-effectiveness threshold is supposed to measure.

In principle, measuring the level at which NICE's threshold should be set would require a complex empirical research project, estimating the cost of a QALY in the various areas of treatment and prevention currently covered by the NHS. When NICE was created in 1999, however, it had no evidence suitable for making such an estimate. Therefore, although the mandate it had received from the Secretary of State required NICE to take the ratio of costs to benefits of heath technologies into account and NICE's committees asked producers to provide £/QALY estimates for such technologies whenever they were available, NICE worked for a while without any formal £/QALY threshold (NICE 2001). Indeed, in the first few years of NICE's life, its representatives were often at pains to point out that no £/QALY threshold was used by the Institute to issue recommendations.[6]

A couple of years after NICE's inception, both external observers and actors from within NICE started looking back at the decisions that it had taken thus far and for which £/QALY estimates were available to establish whether decisions were aligned with the cost effectiveness of the technologies under appraisal. In 2001, James Raftery (2001) found that by restricting the use of technologies to specific subgroups of patients, NICE had kept the cost per QALY of all recommended interventions below £30,000, apart from a single exception. Also in 2001, Sir Michael Rawlins, NICE's then chairman, identified the same upper limit beyond which NICE had been reluctant to issue positive recommendations.[7] Within a few months, it became evident that a single cutoff point was not enough to explain well the decisions that NICE had been making, and in 2002, both £20,000/QALY and £30,000/QALY were highlighted as important figures. Indeed, rejections had been extremely sparse below the £20,000/QALY mark, while the likelihood of rejection appeared to increase beyond it and a positive recommendation became very unlikely above £30,000/QALY (Towse and Pritchard 2002).

The message sent by these early historical analyses of NICE's decision pattern was that NICE had issued recommendations *as if* it had the cost-effectiveness threshold that it lacked the evidence for (although the threshold looked like a "soft" threshold, centered on a range of values, as opposed to a single cutoff point). NICE's reaction to this message is striking: it made the £20,000–£30,000/QALY range into the threshold that NICE committees would be expected to follow in future appraisals. NICE's threshold is supposed to be a measure of an opportunity cost. This supposition means that the level at which the threshold should be placed is a matter of empirical analysis given the cost effectiveness of currently funded health care areas. Identifying a £/QALY threshold by looking at the pattern of past HTA decisions made in absence of any such analysis looks like pulling oneself up by one's own bootstraps. However, this is what NICE did. Michael Rawlins and Anthony Culyer from NICE explained in a 2004 article that a review of past decisions grounded the threshold that NICE would use in the future. The £20,000–£30,000/QALY threshold was enshrined in the 2004 edition of NICE's *Guide to the Methods of Technology Appraisal* and is still used today (NICE 2004; Rawlins and Culyer 2004). Yet this seems a remarkably haphazard way in which to measure such a socially and politically sensitive value.

Some might object that the process through which NICE identified its threshold looks problematic only if we endorse an overly simplified account of what good measurement involves—an account according to which the very first attempt at measurement must fully satisfy the theory of the object to be measured and what it takes to measure it. Could not NICE's setting of the threshold at £20,000–£30,000/QALY be interpreted as a first and admittedly imperfect iteration of a process of measurement of the relevant opportunity costs that would get better at approximating the relevant theory over time?[8] We are doubtful about this interpretation for at least two reasons.

First, even though the first iteration of a process of measurement does not need to fully satisfy the theory of the object to be measured, it seems plausible to require that it should have at least *some* link to such theory, and this link appears to be missing in NICE's original setting of the threshold. The pattern of NICE's past decisions is simply irrelevant to what the threshold is supposed to measure if past decisions are not grounded in an empirical analysis of the cost of a QALY in different areas of NHS expenditure. NICE has always been open about the fact that its threshold figures are grounded in no such analysis. For example, Rawlins commented during an interview that "[t]he £30,000 emerged. I've always said that it's not locked in some empirical basis. It emerged. And it emerged during the first year or two of the appraisal committee meeting" (Appleby 2016, 161). Second, even if we bracketed the issue of how NICE originally arrived at the £20,000–£30,000/QALY figure, NICE's behavior *after* it endorsed the figure does not fit well with the picture of the right sort of iterative process (i.e., one where, iteration after iteration, the threshold is based on a closer approximation of the cost effectiveness of currently funded technologies and therefore better approximates the relevant theory of the object to be measured).[9]

In a 2009 workshop on the threshold, NICE found some reassurance in the outputs of a research project carried out by Peter Smith and other economists of the University of York. Although based on rather limited data, this study was interpreted as suggesting that "NICE is probably not completely out of line in using its current £20–30K per QALY" (NICE 2009).[10] However, four years later, a major study was published that built on this older research project with damning implications for NICE. In a nutshell, Karl Claxton and the other authors of this study aimed to provide an estimate for NICE's threshold based on an empirical analysis of the decrease in the spending of a local commissioning authority that leads to the loss of one QALY's worth of health gain through displacement. The estimate that Claxton and colleagues came up with for NICE's threshold is slightly lower than £13,000/QALY—less than half of the £30,000/QALY figure and far away from even the lower end of NICE's current range (Claxton et al. 2013)! Obviously, this study has disastrous implications for the ability of the current threshold to indicate when the approval of a new technology does more harm than good in terms of health benefits. Therefore, it seems that unless it had some objection to its reliability, NICE should amend its traditional £20,000–£30,000/QALY threshold in light of the new evidence.

NICE did not voice any objection against the evidence put forward by Claxton and colleagues. Still, it rejected the implication that NICE's threshold should be corrected. In explaining this choice, NICE's chief executive Sir Andrew Dillon appealed to considerations that bore no relation to the theory of opportunity costs that lies behind NICE's attention to the cost effectiveness of new technologies. Indeed, the problem with a £13,000/QALY threshold was that it "would mean the NHS closing the door on most new treatments," therefore failing to provide incentives for the pharmaceutical industry to bring about innovation (Dillon 2015). This revelation is a fitting last chapter in the complicated history of NICE's threshold so far. But how should this history be evaluated? This is the question we wish to tackle in the next section.

JUSTIFYING THE THRESHOLD?

As a measure of opportunity costs, the initial calculation of the £20,000–£30,000/QALY threshold seems bad, and the apparent refusal to change this threshold in light of later evidence even worse: using this number is, in effect, systematically to fund treatments that given NICE's stated aims, it should not be recommending for funding (and, thereby presumably, not leaving sufficient funds to fund drugs that should be funded). However, there is a different way of understanding this case, as reflecting the complex relationship between "measures" and "standards," which while not necessarily endorsing NICE's initial calculation and current apparent intransigence, may complicate our assessment.

A "standard" is a conventional rule, which specifies conditions that must be met before something else can or should happen. Often, standards employ numerical terms, as a way of translating between numerical measures and action. For example, within many UK higher education institutions, achieving 70 percent on an exam is necessary and sufficient for achieving a First-Class degree; this "standard" allows us to translate from numerical measurements of candidates' performance to their final degree classification. When standards are numerical, the relevant numbers can be either "fixed," as in the example above, or "floating." As an example of the latter, we might adopt a different standard for a First-Class degree: that the candidate achieves whatever mark results in 20 percent of candidates achieving a First. Our numerical standard for a First might then "float": 68 percent one year and 71 percent another. An interesting feature of NICE's work, quite apart from the odd process through which NICE originally arrived at the £20,000–£30,000/QALY figure and the response to the Claxton critique, is that given its stated aims, the £20,000–£30,000/QALY threshold should "float," but instead it is "fixed." In the section "The Threshold: The Theory," we explained that given that the threshold is supposed to reflect opportunity costs, it should change as the NHS budget changes, as new health needs emerge, technologies are introduced, and so on; however, NICE never varies—and seems to lack any mechanism for varying—its threshold.

In the subsection below, "From an Argument for Fixed Standards to the Setting of NICE's Threshold," we will first sketch an argument for thinking that a "fixed" standard may be justifiable and then consider the relevance of this argument for understanding the original setting of NICE's threshold. In the next subsection, "Can Anything Be Said in Favor of NICE's Response to Claxton et al.?" we will turn to investigating the relevance of our arguments to NICE's response to the Claxton critique.

From an Argument for Fixed Standards to the Setting of NICE's Threshold

Constantly updating the standard—using a "floating" threshold—would be extremely complex from a technical perspective. However, this complexity is, presumably, only part of the story why NICE prefers a fixed standard. Rather, the preference for a fixed standard seems related to a more fundamental sociological dynamic, explored well by writers such as Ted Porter. This dynamic concerns how bureaucratic needs shape systems of measurement and assessment. As Porter explains, in a "public measurement system," such as the systems used by state bureaucracies (e.g., NICE or the NHS), there are strong forces requiring "standardization" (i.e., that like cases be treated alike or, at least, be seen to be treated alike) and "proper surveillance." As a result, "there is a strong incentive to prefer readily standardizable measures to highly accurate ones, where these ideals are in conflict" (Porter 1994, 391). For example, state-mandated systems for measuring the toxicity of chemicals may often differ from the "most accurate" measures. Indeed, Porter goes further, suggesting that "if an eccentric manufacturer were to invest extra resources to perform a state-of-the-art analysis, this would be viewed by the regulators as a vexing source of inter-laboratory bias, and very likely an effort to get more favorable measures by evading the usual protocol, not as a welcome improvement in accuracy" (Porter 1994, 391). Of course, Porter is describing systems for measuring and assessing quantities relevant to some standard—rather than what we have called "standards" themselves—but it is easy to apply his model to NICE. A "floating" measure of the relevant cost per QALY would, given NICE's stated aims, be more apt than a "fixed" measure as part of the standard for funding decisions. However, just as using a state-of-the-art toxicity measure would be bureaucratically complex, the same might be true of using a floating standard even though they are both more "accurate."

The actions of a state body can be analyzed and assessed from many viewpoints. Clearly, one central question we can and should ask of state action is whether it abides by basic ethical norms of fairness or accountability. Therefore, to translate key themes of Porter's sociological analysis into the basis for an ethical assessment, we wish to ask, what might justify a fixed standard? There are two reasons to prefer such a standard. First, a fixed standard avoids potential problems of (perceived) diachronic fairness. Use of a floating standard would, presumably, imply that NICE should deny drugs to some patients who would, a year previously, have seen the same drugs approved (or conversely, that NICE should recommend drugs on the basis of their cost effectiveness while equally cost-effective drugs were rejected because of their cost per QALY the previous year). Such decisions might, in fact, be justifiable—given the underlying logic of cost-effectiveness analysis—but they would at least appear (and arguably be) unfair in that they would treat identical health technologies differently.[11] Second, changing the threshold every few months would make it more difficult for outsiders to assess and discuss NICE's decision making procedures. In turn, this would have a detrimental impact on goods such as transparency and democratic oversight.[12] What is the value of (perceived) fairness, transparency, and democratic oversight? This is an open question, but given NICE's regulatory role, such goods are not to be given up lightly. Therefore, even if NICE's use of a fixed standard eschews accurate measurement of the constantly varying "true" opportunity cost, it is not clear that a more accurate floating standard would, all things considered, be preferable.

This discussion of the justifiability of a fixed threshold highlights a set of goods that are external to the narrow logic of cost-effectiveness analysis but that can be used as a counterbalance to it. In turn, consideration of these goods sheds light on the justifiability of a more fundamental choice that NICE made—that of having an *explicit* threshold at all and, therefore, setting the £20,000–£30,000/QALY figure despite the lack of solid empirical evidence concerning opportunity costs. There are three possible arguments in favor of NICE's setting an explicit standard. To start with, the very existence of an explicit threshold appears to foster transparency and democratic oversight at an even deeper level than the choice of a fixed over a floating threshold. Furthermore, an explicit threshold ensures coordination across NICE's five health technology appraisal committees. By standardizing the process through which those committees are supposed to issue recommendations about drugs, an explicit threshold reduces the risk of an unfair "committee lottery," or, in other words, the risk that patients in need of a new drug will have that drug denied simply because it has been evaluated by one NICE HTA committee and not another. Like the diachronic fairness fostered by a threshold that is not constantly recalculated, standardization across NICE's various HTA committees makes it more probable that like cases will be treated alike by decision makers. Finally, an explicit threshold sends a clear message to the pharmaceutical industry. By stating that NICE is unwilling to recommend spending more than a certain amount of money for a QALY, an explicit threshold places pressure on industry to lower the price of many expensive drugs, ultimately benefiting the NHS.[13]

Even if we can justify using a measurement estimate that is both explicit and fixed, it seems important that the estimate be roughly correct. There may be good reasons to use a blood pressure reading of 140/90 as a point for prescribing cardiovascular medications for all patients even if a more complex standard would provide more targeted care; the "benefits" of such a system (in terms of ease of implementation, transparency, ease of communication) might outweigh the "costs" (in terms of occasional over- or underprescription). However, such arguments require that 140/90 tracks something like "actual" increased risk; if it doesn't, ease of use seems irrelevant. Similarly, we might worry that even if there are good reasons for NICE to use an explicit and fixed standard, those arguments are not worth much if that figure wildly underestimates the opportunity costs of new drugs. We will now suggest a way of challenging this thought.

So far, we have used Porter's sociological model to explore one way in which the "constraints" on administrative agencies *may* justify a preference for measures that given those agencies' stated roles, seem peculiar (e.g., a fixed over a floating standard). However, regulatory agencies may serve multiple roles, some of which may differ from their stated roles, and these additional roles may, themselves, be morally or politically valuable. To make this clearer, consider the complex political backdrop to NICE's work. Decisions about drug funding are, of course, highly controversial. There are good reasons for politicians and governments to want such decisions to be removed—at least in part—from their (perceived) sphere of influence. One function of NICE is to serve this end (Gash et al. 2010, 18). From this perspective, it is important that NICE's decisions be seen as impersonal. Political controversy can be avoided if decisions are seen as the result, at least in part, of "objective" number crunching. In the words of Rottenberg and Merry, numeric representation in governance achieves the political purpose of demonstrating, "adherence to public responsibility and absence of personal or group bias"

(2015, 8). It might be added that numbers carry immense symbolic authority as guarantors of objectivity, rigor, and universality and hence may contribute to institutional legitimacy quite independently of their precision and accuracy (Sauder and Espeland 2015, 436). To place these comments in a broader context, we might say that one of NICE's key *political* functions is to ensure that funding decisions are—or are perceived to be—"procedurally objective," in the sense that they are determined by application of impersonal rules rather than individual idiosyncrasies; in Megill's nice phrase, they are "untouched by human hands" (Megill 1994, 10). We can contrast this sense of "objectivity" with the "absolute" sense of "objectivity" at play, for example, when we describe measures as "representing reality as it really is," as in the case at hand, when we ask whether NICE's threshold "really" reflects the "true" opportunity costs of NHS spending (Megill 1994, 1).

Building on these comments and moving once again from a more sociological to a more philosophical level of analysis, we can distinguish two functions of the threshold: one is the "stated aim" (i.e., to ensure that concerns about cost effectiveness are accurately reflected in decisions about resource allocation). As explained above in "The Threshold: The Theory," there is a familiar debate in the philosophical literature over how best to balance the aggregative and maximizing logic of cost effectiveness with more egalitarian or other distributive considerations. From the perspective of such debates, cost effectiveness is a morally relevant consideration, and therefore NICE should revise its threshold to £13,000 and maybe even allow it to "float" around this point. (Strictly speaking, this conclusion might be sensitive to the moral weight we assign to cost effectiveness, but the underlying issue is clear: £30,000/QALY must be rejected!)

However, it is not clear that NICE's only normatively relevant function is to act as a kind of massive central planner (even a central planner whose decisions are to be guided by *more* than cost effectiveness). Rather, we might understand its role differently (i.e., ensuring consistency across cases, placing limits on political pandering to electorally significant groups, allowing for rational planning, stabilizing drug prices, ensuring that decisions can be assessed and criticized, creating a broad democratic debate over the ethics of NHS resource allocation, and so on). From an ethical perspective, these are all potentially important goods, which require only that NICE's recommendations are procedurally objective. The precise numbers specified in these procedures are a bit like the rule that football teams have eleven players on each side. The number eleven is not magical, in that one could play a sport very like football with twelve people, but we need to settle on some number if there is to be fair competition, if teams are to be able to plan strategy, and so on. What is required is that there be *some* defensible number, not that the number reflects some fact, such as that football is "best played" as eleven a side.

Our claim, then, is that when we think about NICE's work through the prism of "procedural objectivity" and of the politically and morally important goods that this sort of objectivity generates, it seems less important that the number it chose was "true" than that it chose some broadly acceptable number at all and then set this number as a kind of explicit benchmark for itself and others to follow. Clearly, given that there was some kind of apparent implicit agreement on the £20,000–£30,000/QALY figure, choice of this number was not completely unreasonable for this purpose. Admittedly, there is something odd about this approach, in that a number that looks like—and is ostensibly described and justified as—a measure turns out to function more like a convention. We return to this issue below, but note here that any serious attempt to think about measurement and standards in institutional and political contexts, such as NICE, that is not alert to issues of coordination and fairness is likely to overlook morally and politically important concerns.

Can Anything Be Said in Favor of NICE's Response to Claxton et al.?

Once we stress NICE's coordinating role, concerns about the initial choice of the £20,000–£30,000/QALY threshold are less pressing even though at that time, there was no empirical evidence connecting that figure to "true" opportunity costs. Still, you might think that NICE's stated goal—to ensure that money is spent in a cost-effective manner—is of great importance. From this perspective, now that empirical evidence is available, NICE should change its threshold. Doing so may seem compatible with serving its other politically and morally relevant functions: after all, we can just as well coordinate around £13,000/QALY as £20,000–£30,000/QALY.

One might explain NICE's response to the work of Claxton et al. as an instance of a more general familiar sociological phenomenon of bureaucratic inertia: it would be tiresome and costly to change the threshold. Furthermore, it seems there is little political impetus to do so (plausibly, matters would be very different were the report to have suggested a *higher* threshold; there are many patient advocacy groups who would agitate for a higher threshold). Still, important and interesting as these dynamics are, it is hard to see how they might justify at a philosophical level NICE's apparent insouciance in the face of Claxton's critique.

However, the model developed above that viewed NICE as serving many political functions beyond its "official" role provides a more nuanced assessment of NICE's refusal to shift its threshold. Consider again Andrew Dillon's justification for retaining the higher threshold quoted at the end of the previous section: that a change would disincentivize innovation. His argument seems wrongheaded if we view NICE's work solely as a central planner maximizing under certain distributive constraints the health benefits that NHS interventions can produce. But in our model in which NICE serves many different political functions, we might plausibly say that one function of NICE is to promote pharmaceutical innovation and, hence, contribute to the British economy. If this function is viewed as normatively valuable and if it is true that changes to the threshold would negatively affect innovation and, hence, the economy, then maybe there is some argument for retaining the deeply flawed threshold. Furthermore, it may be possible to argue that if NICE is supposed to view the pharmaceutical industry as a kind of stakeholder in its work, then, given the long-term nature of planning in the pharmaceutical industry, there may be considerations of fairness that count against a rapid change in the threshold's value. In making these remarks, we are not endorsing such arguments. It is unclear that NICE should have the role of promoting innovation and unclear that a lower threshold would stifle—as opposed to incentivize—drug development. What we are suggesting, rather, is that no proper assessment of NICE's work can go forward without proper attention to the purposes behind its measures (or the purposes they have come to serve). Our approach allows us to engage with Dillon's justifications rather than treat them as necessarily irrelevant.

When we have some politically mandated system of measurement or standard that incorporates some numerical value, we can always ask whether that system or standard is fit for purpose. When making such an assessment, we might be willing to sacrifice a certain degree of accuracy for other goods. For example, if we assume that NICE's threshold is intended to capture concerns about cost effectiveness, we can ask whether the £20,000–£30,000/QALY threshold is fit for that purpose. When we realize that it should, but does not, float, we might be willing to tolerate this "inaccuracy" as the "price" for, say, ease of use. However, we also made a second, stronger claim: that NICE's threshold serves multiple functions not related to cost effectiveness but rather to the appearance of fairness, to stabilizing expectations, to facilitating democratic deliberation, and so on. From this broader perspective, "fitness to purpose" is more complicated because what matters is not only that the numerical standard accurately reflects opportunity costs but that there is a standard that remains stable across time and maintaining those goods may require defending this number even when it is "wrong" in the narrower sense. None of these arguments straightforwardly justifies NICE's decisions. After all, NICE's scheme turns out not to incorporate a concern that many do think is important: whether drugs are "cost effective." However, any critique of NICE's inertia must start from an assumption about its proper normative function; there is no point in measuring opportunity costs more or less accurately if those costs are irrelevant to NICE's work. Once this point is made explicit, it is entirely proper to ask whether NICE serves other proper normative functions and if so, whether these functions might, at least in part, justify the (apparently arbitrary) threshold bequeathed by historical accident. To ignore these issues is to endorse one set of value concerns as the only ones that are proper without adequately considering all the alternatives.

CONCLUSION

Many features of the £20,000–£30,000/QALY threshold, such as how it was derived, that it is "fixed," and more, make little sense considering the standard account of NICE's purposes, including the accounts NICE itself gives. Whatever else we can say about these numbers, they are not a good measure of the opportunity costs of funding new technologies. However, many of these puzzling features make sense, and may even be justified, when we rethink NICE's functions not only as a central planner but also as a guarantor of procedural objectivity in a domain of deep conflict. Clearly, this story has implications for our understanding of the work of NICE, similar HTA bodies in other jurisdictions, and heated debates over the "proper" way of measuring and rationing health care interventions more generally. However, it also has a broader implication for measurement in (and maybe beyond) medicine. Any measure of cost effectiveness is, in a trivial sense, value laden because, for example, we need to choose what effects to measure. To measure the effectiveness of a treatment along some dimension is, if only implicitly, to assume that this dimension is of prudential, moral or political significance.

These are familiar claims, well covered in the now extensive literature on how to construct measures of health-related quality of life. What is less obvious, but no less important, is that the construction of measures to be used for policy making may be subject to further moral and political considerations that may be in tension with the aim of accurately representing the aspect of reality that these measures are supposed to track. Demands that users of measures be accountable to others for their decisions, for example, may give us reasons to prefer measures that have a certain sort of "inflexibility," for example and thus do not always track what we seek to measure. Furthermore, measures may take on a "life of their own," such that claims that measures are inadequate guides to underlying phenomena may fail to consider the role that these measures play in the complex ecology of policy. For example, when a putative measure becomes a standard or a target, we need not only ask how well it functions as a measure but also what the consequences are of its further uses. Taking account of such concerns is not to say that questions of accuracy are irrelevant to our assessment of measures. Rather, it is to add a twist to the truisms that measurement is always for a purpose and that adequacy is relative to our purposes: that the purposes of measurement are often multiple and might even be opaque. To make such claims is not to dismiss a measure but rather to open up a new question, about whether those purposes are sufficiently valuable that we should tolerate inaccurate measures.

NOTES

1. For this process in general, see NICE (2008, 17–19) and NICE (2013, 72–74). For the equity weightings, see Rawlins, Barnett, and Stevens (2010).
2. See Bognar and Hirose (2014, 1–6) for a succinct version of these arguments. The first section will cover them in greater detail.
3. For but one recent treatment of this topic from a philosophical perspective, see Hausman (2015).
4. Of course, the reference to the *single* least cost-effective intervention that is currently funded makes sense only on the simplifying assumption that the introduction of the new technology does not have a larger budgetary impact than that intervention currently has. In principle, if the technology under appraisal has a particularly large budgetary impact, the cost-effectiveness threshold should be lowered.
5. For the importance of balancing health maximization and distributive concerns, see Bognar and Hirose (2014, 53–78 and 104–26), Brock (2004), and Daniels and Sabin (2008, 30–34).
6. For example, see the discussion of Michael Rawlins's public statements in Littlejohns (2002, 32).
7. See the references to Rawlins's discussion of the topic at NICE's 2001 annual public meeting in both Littlejohns (2002, 31–32) and Towse and Pritchard (2002, 26–27).
8. See Tal (2013) for an excellent account of why naïve theories of measurement are descriptively and normatively problematic.
9. To use an effective image introduced by Culyer et al. (2007) to outline their idea of NICE's role, we aim to show that NICE has been a poor "threshold-searcher" also after endorsing the £20,000–£30,000/QALY figure in 2004.
10. For more on the research project under discussion, see Martin, Rice, and Smith (2009).
11. That like cases should be treated alike is often proposed as a basic principle of fairness or justice in the allocation of health care resources. For example, see Clark and Weale (2012, 306–307) and Daniels and Sabin (2008, 47–49).
12. To cite but one influential account of fair procedures, transparency is one of the four conditions defining a fair process for the allocation of health care resources according to Daniels and Sabin (2008, 43–66). Also, Daniels and Sabin (2008, 59–60) argue that part of the value of fair procedures in health care resource allocation is that such procedures foster democratic deliberation in society at large.
13. On the other hand, however, there are cases in which a fixed threshold that operates as a standard might just as easily have the opposite effect, namely, encourage industry to come in at the higher end in its pricing if it still comes in within the approved range. This concern is encompassed by the observation of scholars of policy and audit that standards are also susceptible to treatment as targets, which can have

perverse effects with respect to the goals that might drive their use in the first place. For this point in general, see Shore and Wright (2015, 425). Bevan and Hood (2006) provide a more focused analysis of the gaming of targets in the NHS.

REFERENCES

Appleby, John. 2016. "What's In and What's Out? The Thorny Issue of the Threshold." In *A Terrible Beauty: A Short History of NICE*, edited by Nicholas Timmins, Michael Rawins, and John Appleby, 154–69. Nonthaburi, Thailand: HITAP.

Bevan, Gwyn, and Christopher Hood. 2006. "What's Measured Is What Matters: Targets and Gaming in the English Public Health Care System." *Public Administration* 84: 517–38.

Bognar, Greg, and Iwao Hirose. 2014. *The Ethics of Health Care Rationing*. New York: Routledge.

Boseley, Sarah. 2015. "Postcode Lottery for Cancer Drug as Nice Rules Kadcyla Too Expensive." *The Guardian*, November 27. Accessed July 9, 2016. https://www.theguardian.com/society/2015/nov/17/postcode-lottery-cancer-drug-nice-rules-kadcyla-too-expensive.

Brock, Daniel. 2004. "Ethical Issues in the Use of Cost Effectiveness Analysis for the Prioritisation of Health Care Resources." In *Public Health, Ethics, and Equity*, edited by Sudhir Anand, Fabienne Peter, and Amartya Sen, 201–23. Oxford and New York: Oxford University Press.

Clark, Sarah, and Albert Weale. 2012. "Social Values in Health Priority Setting: A Conceptual Framework." *Journal of Health Organisation and Management* 26: 293–316.

Claxton, Karl, Stephen Martin, Marta Soares, Nigel Rice, Eldon Spackman, Sebastian Hinde, Nancy Devlin, Peter C. Smith, and Mark Sculpher. 2013. "Methods for the Estimation of the NICE Cost Effectiveness Threshold." *CHE Research Paper 81*.

Culyer, Anthony, Christopher McCabe, Andrew Briggs, Karl Clazton, Martin Buxton, Ron Akehurst, Mark Sculpher, and John Brazier. 2007. "Searching for a Threshold, Not Setting One: The Role of the National Institute for Health and Clinical Excellence." *Journal of Health Services Research and Policy* 12: 56–58.

Daniels, Norman, and James Sabin. 2008. *Setting Limits Fairly: Learning to Share Resources for Health*. 2nd Edition. Oxford and New York: Oxford University Press.

Dillon, Andrew. 2015. "Carrying NICE over the Threshold." *NICE Blog*, February 19. Accessed July 9, 2016. https://www.nice.org.uk/news/blog/carrying-nice-over-the-threshold.

Gash, Tom, Jill Rutter, Ian Magee, and Nicole Smith. 2010. *Read before Burning: Arm's Length Government for a New Organisation*. London: Institute for Government.

Hausman, Daniel. 2015. *Valuing Health: Well-Being, Freedom, and Suffering*. Oxford: Oxford University Press.

Littlejohns, Peter. 2002. "Does NICE Have a Threshold? A Response." In *Cost Effectiveness Thresholds: Economic and Ethical Issues*, edited by Adrian Towse and Clive Pritchard, 31–37. London: King's Fund.

Martin, Stephen, Nigel Rice, and Peter Smith. 2009. *The Link between Healthcare Spending and Health Outcomes for the New English Primary Care Trusts*. London: Health Foundation.

McCabe, Cristopher, Karl Claxton, and Anthony Culyer. 2008. "The NICE Cost-Effectiveness Threshold: What It Is and What That Means." *Pharmacoeconomics* 26: 733–44.

Megill, Allan. 1994. "Introduction: Four Senses of Objectivity." In *Rethinking Objectivity*, edited by Allan Megill, 1–20. Durham: Duke.

NICE. 2001. *Guide to the Methods of Technology Appraisal 2001*. London: NICE.

NICE. 2004. *Guide to the Methods of Technology Appraisal 2004*. London: NICE.

NICE. 2008. *Social Value Judgements: Principles for the Development of NICE Guidance*. London: NICE.

NICE. 2009. *Threshold Workshop: Report of a Technical Meeting Organised by NICE*. London: NICE.

NICE. 2013. *Guide to the Methods of Technology Appraisal 2013*. London: NICE.

NICE. 2015. "Kadcyla Price Too High for Routine NHS Funding, Says NICE in Final Guidance." Accessed July 9, 2016. https://www.nice.org.uk/news/press-and-media/kadcyla-price-too-high-for-routine-nhs-funding-says-nice-in-final-guidance.

Porter, Theodore. 1994. "Making Things Quantitative." *Science in Context* 7: 389–407.

Raftery, James. 2001. "NICE: Faster Access to Modern Treatments? Analysis of Guidance on Health Technologies." *British Medical Journal* 323: 1300–1303.

Rawlins, Michael, David Barnett, and Andrew Stevens. 2010. "Pharmacoeconomics: NICE's Approach to Decision-Making." *British Journal of Clinical Pharmacology* 70: 346–49.

Rawlins, Michael, and Anthony Culyer. 2004. "National Institute for Clinical Excellence and Its Value Judgments." *British Medical Journal* 329: 224–27.

Rottenberg, Richard, and Sally Engle Merry. 2015. "A World of Indicators: The Making of Governmental Knowledge through Quantification." In *The World of Indicators: The Making of Governmental Knowledge through Quantification*, edited by Richard Rottenberg, Sally E. Merry, Sung-Joon Park, and Johanna Mugler, 1–33. Cambridge: Cambridge University Press.

Sauder, Michael, and Wendy Espeland. 2015. "Comment on 'Audit Culture Revisited: Rankings, Ratings, and the Reassembling of Society.'" *Current Anthropology* 56: 436–37.

Shore, Chris, and Susan Wright. 2015. "Audit Culture Revisited: Rankings, Ratings, and the Reassembling of Society." *Current Anthropology* 56: 421–44.

Smith, Richard, and Jeff Richardson. 2005. "Can We Estimate the 'Social' Value of a QALY? Four Core Issues to Resolve." *Health Policy* 74: 77–84.

Tal, Eran. 2013. "Old and New Problems in Philosophy of Measurement." *Philosophy Compass* 8: 1159–73.

Towse, Adrian, and Clive Pritchard. 2002. "Does NICE Have a Threshold? An External View." In *Cost Effectiveness Thresholds: Economic and Ethical Issues*, edited by Adrian Towse and Clive Pritchard, 25–30. London: King's Fund.

Chapter 11

Cost Effectiveness

Finding Our Way through the Ethical Morass

Daniel M. Hausman

Distributing health care resources so they bring about the greatest health improvement, which is what choosing the most cost-effective intervention accomplishes, seems initially to be a sensible thing to do. But there are well-known ethical objections. This essay addresses a few of the many difficulties. It asks (1) whether considerations other than the promotion of health should govern the allocation of health care resources and how much weight they should have, (2) whether ethical problems with the application of cost-effectiveness ratios should be addressed by adjusting the effectiveness measure to include ethical considerations, and (3) whether *fairness* demands that we should place a greater weight on treating those who are more seriously ill.[1]

FOUR QUALMS CONCERNING RATIONING BY COST EFFECTIVENESS

If there is some way to measure both the improvement in health that alternative policies will bring about and the costs of the policies, one can determine the ratio of the cost to the improvement for each policy. For example, Gilead Sciences is charging $84,000 for a course of its new drug, Sovaldi. Although very expensive, Solvaldi cures hepatitis C and thus provides a very substantial health benefit. Measuring that benefit in a unit, such as Quality-Adjusted Life Years, or QALYs, one can specify the cost-effectiveness ratio, that is the dollars per QALY of a treatment with Sovaldi.[2] If one devotes a fixed health budget to the most cost-effective interventions, one will then maximize the health benefits of the health budget. If that is what the health ministry aims to do, then the health ministry should rely on cost-effectiveness information to distribute its budget.

Although the argument for employing cost-effectiveness information to allocate the health budget is straightforward and plausible, there are serious ethical objections to the objective of maximizing health and the employment of cost-effectiveness information. For a classic discussion of these objections, see Brock (2003). In this essay, I shall comment on the following four objections.

1. Should the objective of the health department be to promote the health of the nation's population or to promote its welfare?

Health interventions have consequences other than their health benefits. Their costs are forgone opportunities to provide other benefits, which might, ironically, promote health more effectively than health policies. It is well known, for example, that education makes a huge contribution to health. On the other hand, the benefits of health policies extend beyond their immediate consequences for health. Health improvements in the working-age population increase productivity and enhance the society's wealth. Curing illnesses and lessening disabilities diminish the demands on families and other caregivers. Moreover, quite apart from their consequences for health, health policies may have large effects on people's well-being. Whether or not universal health care improves health, it enhances financial security. Whether the "effectiveness" in cost-effectiveness ratios should be effectiveness at improving health is a serious issue, and one of the issues this essay will address. This is less an objection to the use of cost-effectiveness ratios to allocate health-related resources than an objection to the conceptualization and hence the measure of "effectiveness." Call this "the well-being objection."

2. Should those whom it is less cost effective to treat get no treatment?

Treatments for some health conditions are more cost effective than treatments for others, and it may also be more cost effective to treat some people than others even though they have the same health problem. For example, untreated hypertension is a much more serious problem among the disadvantaged in the United States than among the relatively affluent. But it is more expensive to identify those among the disadvantaged who need treatment, to subsidize their treatment, and to follow up to secure compliance with the treatment. It may thus be the case that it is more cost effective to screen for hypertension exclusively among the middle and upper classes. But such a use of resources would appear to be unfair. Even though it is less cost effective to treat Jill rather than Jack, if they have the same health problem, they should both be treated or, if resources are scarce, they should both have a chance—indeed perhaps an equal chance—to be treated. This objection to the use of cost-effectiveness information is known in the literature as the "fair-chances objection."

3. Should health-related resources be distributed in the most cost-effective way when doing so aggravates inequalities?

Those who are more expensive to treat or who will not benefit as much from treatment may be members of groups that have been unfairly treated or who suffer from exclusion and unjust discrimination. Moreover, those who have disabilities may achieve smaller health improvements from treatments than those who are not disabled. If 1.0 is the measure assigned to full health, then a medical intervention that gives someone in full health an additional year of life provides him or her with one QALY. Since the measure of health assigned to disabilities will be less than 1.0, giving someone with a disability an additional year of life provides him or her with less than one QALY. If lifesaving resources are in short supply, it will be cost effective to treat those with disabilities only after one has treated everyone who is not disabled. Such a policy would be heinous. Call this the "discrimination objection."

4. Should the priority assigned to treatments depend exclusively on their cost and on the magnitude of the change in health that they cause and not at all on how severely ill someone is?

Suppose that on a scale of 0 to 10, Jack's pain is at level 9, while Jill's pain is at level 4. An analgesic that reduces Jack's pain to level 7 and Jill's pain to level 2 is equally effective for both Jack and Jill; since the cost is the same, it is a matter of indifference from the perspective of cost effectiveness who gets the analgesic. But Jack is in agony, while Jill is just very uncomfortable. In failing to consider the severity of people's pretreatment health states and focusing exclusively on the change that treatments bring about, relying on cost effectiveness to allocate resources seems to be morally obtuse. Call this the "severity objection."

Although these four objections to allocating health-related resources by their cost effectiveness are more than enough for one essay, I should mention that these objections are not exhaustive. Here are some of the additional questions that I shall not discuss:

- *How should one measure effectiveness?* The measure of effectiveness, which rests on the measure of "generic" or overall health, is shot through with ethical questions, and partly as a consequence of disagreements concerning how to answer these questions, different systems of health measurement disagree radically on the numbers they assign to health states. There is a disagreement about whether effectiveness rests on a measure of the quantity or magnitude of health or whether effectiveness depends on a measure of the *value* of health—that is, on a measure of how good or bad someone's health is. In common with most health economists, I shall assume that generic health measures are measures of the value of health.

- *Should an additional year of life in full health count the same for everyone regardless of a person's age?* This decision has huge ramifications because rejecting age weighting winds up prioritizing saving the lives of the very young, who can expect to live for many decades.
- *Should future health benefits be discounted?* Is curing a case of malaria in a five-year-old today of more value than curing a case of malaria in a similar five-year-old ten years from now? Once one distinguishes this from the very different question of whether it is better to cure a five-year-old's malaria today or to let her or him suffer from the disease another ten years before curing it, it is hard to see why the timing of a cure of similar patients should matter. But there is a great deal more to be said about this question.
- *How should one calculate the benefit of saving a life?* The benefit apparently depends on how long the individual will live if treated and the quality of life during the additional years. If one relies on actual life expectancies, saving the life of someone who has a shorter life expectancy owing to previous disadvantage will bring a smaller benefit than saving the life of someone from a more advantaged region or social class. To avoid this, should one assign a uniform life expectancy or perhaps a different life expectancy for men and women? What should that life expectancy be?
- *Should providing several small health improvements ever take priority over saving a life or alleviating an extremely serious health problem?* The original Oregon Medicaid rationing scheme found that capping teeth was more cost effective than performing appendectomies. So if there were not sufficient funds to do both, it would be cost effective to cap teeth and allow those with appendicitis to die. This implication of relying on cost-effectiveness information seems outrageous. Philosophers have responded by arguing that some benefits are too small to be "relevant" to decisions about allocating resources when death or serious health problems are at issue.

These additional objections, which I will not discuss, are obviously well worth discussing. But there is only so much that one essay can accomplish.

CHANGING THE MEASURE OF EFFECTIVENESS FROM HEALTH TO WELL-BEING

Fully addressing the four objections that I plan to discuss, let alone the other objections as well, requires a book, not a chapter. I propose instead to ask narrower questions about how to address the objections. In this section, I will consider whether the relevant measure of effectiveness should be (the value of) health improvement or whether as John Broome (2002) has argued, the relevant measure should instead be improvement in overall well-being.

We've already seen one reason for change: health policies affect individuals in many ways in consequence of and sometimes quite apart from their effects on population health. It seems only reasonable that those other consequences should be considered when assessing alternative health policies. One straightforward way to take these other consequences into account is to assess health policies by their consequences for well-being rather than by their consequences for health.

A second argument for measuring effectiveness in terms of well-being rather than in terms of the value of health is, as John Broome (2002) has pointed out, that the benefits or harms of health states are not separable from other factors that benefit or harm individuals. The same health state may have different consequences for someone's life depending on the technological, cultural, or natural environment and the individual's objectives and interests. So there is no such thing as *the* value of a health state. Broome concludes that rather than aiming to maximize the value of population health, policies should aim to maximize the population's well-being. Although Broome is unquestionably correct to maintain that health benefits or harms are not separable from the benefits and harms caused by other factors and that the same token health state may have very different values depending on other factors, the inseparability does not preclude the possibility of defining an average value of a health state or the value of a health state in some specified "standard" circumstances.

Changing the measure of effectiveness from health to well-being would also help with some of the specific ethical objections to relying on cost-effectiveness information to allocate health-related resources. Since those with disabilities, such as deafness, may have just as high a level of well-being as those without these disabilities, there will be less discrimination if the health ministry allocates health-related resources by their effectiveness at promoting well-being than by their effectiveness at improving health. If having a "fair chance" at receiving some benefit affects people's well-being, then allocating health-related resources by their effectiveness at promoting well-being would be less subject to the fair-chances objection than standard cost-effectiveness analysis. Although shifting the objective to promoting well-being clearly does not address all the ethical objections, it does lessen some, which provides a third reason to favor the shift.

So there are three good reasons to measure effectiveness with respect to well-being rather than with respect to health. Moreover, if the only means of promoting either health or well-being available to the health department or ministry are health policies, then this broadening of concern from health to well-being need not be a radical change. In most cases the only way to promote well-being with public health measures or health care policies will be to promote health, and whatever best promotes health will typically coincide with whatever best promotes well-being.

But the two will not always coincide, and the whole conceptualization of the role of a specific department of government, such as the health ministry, will be changed. The most significant change is, of course, the fact that policy choices will depend on measures of well-being rather than on measures of health. In this paper, I am not discussing measures of health, a subject to which I've devoted a whole book (Hausman 2015). Suffice it to say that these measures are extremely problematic. So it might appear that relying instead on measures of the contributions that health policies make to well-being would be a welcome change.

Such is not the case, however, because measures of well-being are, in fact, more problematic than are measures of health. In standard welfare economics, well-being is measured by preference satisfaction. On the face of it, this is extremely unsatisfactory because obviously, individuals may prefer x to y even though y is better for them. One obvious way in which this may happen is if the individual is ignorant of the properties or consequences of the alternatives or has mistaken beliefs about them. It may also be the case that individuals knowingly choose an alternative that is worse for them but that promotes some end for which they are willing to sacrifice some benefits for themselves.

It would be unflattering to economists to suppose that they are unaware of these problems (although many confuse the claim that people's actions are driven by their own preferences—*the interests of a self*—with the claim that people's actions are directed toward their own well-being—*their self interest*). A charitable interpretation regards economists as supposing that people's actual preferences are a rough and ready guide to what their preferences would be if people were fully informed, self-interested, and good judges of their interests who are free of the many rational flaws to which real people are subject. If one supposes that people are good judges of what promotes their well-being, knowledgeable about the consequences and properties of the alternatives, and concerned exclusively about their own well-being, then what they prefer should coincide with what is good for them.

But obviously, people often lack information or have false beliefs. They are not always concerned with only their own well-being. They are subject to many cognitive limitations, and they may not be good judges of what is good for them. So measuring well-being by people's preferences has many drawbacks. Moreover, it faces several problems that measures of the value of overall health avoid.

1. Although there are many ways to be in bad health, there is a single conventional notion of "fully healthy." On the other hand, what constitutes a good life for one person may be utterly different from what makes life good for another. It is only a slight exaggeration to maintain that there is one way to be healthy, which generic health measures attempt to quantify,

while the measurement of well-being must cope with the many ways to have a good life.
2. One reason well-being is diverse while good health is uniform is that what is good for me depends heavily on my goals and values. People's goals and values influence their assessment of their health too but usually to a much lesser extent.
3. Interpersonal comparisons of well-being pose well-known problems,[3] while interpersonal health comparisons are not appreciably more difficult than intrapersonal comparisons. Unlike a single individual's activities, whose values can often be compared with respect to the individual's unchanged aims, and unlike the mental states of an individual, which are experienced by a single subject, the contributions to well-being of the activities of individuals and the quality of their mental states are hard to compare. In contrast, comparisons of which of two people is in better health, like comparisons of whether a person at one time is in better health than that same person at another time, are comparisons of the health states and their trajectories. *Who* occupies those health states is typically irrelevant.
4. Assessments of well-being during some period of an individual's life often depend on what his or her life is like before or after. Our primary object of interest when considering well-being is the individual's whole life.[4] On the other hand, when considering health, we think mainly about how good someone's health is during some period—how healthy someone is now or how healthy he or she was as a child.

These contrasts do not imply that well-being cannot be measured, but they should give one pause before substituting the promotion of well-being as an objective for the health ministry for the objective of promoting health. I think that the concerns that suggested measuring effectiveness with respect to well-being rather than with respect to health, which are valid and important, can be met without changing the criterion in terms of which effectiveness is to be measured. Assigning to the health ministry the task of allocating its fixed health budget in whatever way will maximize the contribution that health resources make to well-being is to demand too much of it. The task is not well defined unless the health ministry is able to coordinate with other branches of government, and the measurement of the consequences for well-being is beyond the capacities of health economists. The proposal largely defeats the purpose of the administrative division of labor, which is designed to simplify the decision making within specific government agencies by more narrowly specifying the objectives to be accomplished. Moreover, a given health improvement usually implies a greater improvement in well-being among those who are otherwise living well. Consequently, focusing on well-being rather than health would aggravate the discriminatory implications of relying on cost effectiveness to allocate health-related resources.

The nonhealth consequences of health policies are important, as are the health consequences of nonhealth policies concerning education, transportation, housing, nutrition, or the environment. But the way to cope with these interactions is not to assign each administrative division the same goal to be promoted through the different means each division has on hand. The health department should aim to improve health, and the education department should aim to improve education. Coping with the interactions between the activities of the different departments is a separate administrative or legislative task that should be the responsibility of a separate agency that determines the budgets and guidelines constraining specific departments.

CHANGING THE MEASURE OF EFFECTIVENESS FROM HEALTH TO THE "SOCIAL VALUE" OF HEALTH

Among health economists, Erik Nord (1999; Nord et al. 1999) has shown a special sensitivity to the apparent ethical failings of relying on cost-effectiveness ratios to allocate health-related resources. Rather than laying out his own normative framework, he has posed the ethical issues as a conflict between the implications of cost-effectiveness considerations and the attitudes of relevant populations and has accordingly relied heavily on surveys of population attitudes in criticizing cost effectiveness and in proposing an alternative. The survey results that have been of special interest to Nord conform to the ethical objections I sketched above in "Changing the Measure of Effectiveness from Health to Well-Being." Although there are many disagreements in the responses to survey questions, large numbers of respondents in several countries apparently endorse the following three principles, which are inconsistent with rationing by cost effectiveness and reflect concerns about severity, fair chances, and discrimination:

1. Survey respondents prioritize treating those who are more severely ill even when the health benefits are the same or somewhat less.
2. Survey respondents give more weight to treating those who have a lesser capacity to benefit from treatment than is consistent with cost effectiveness.
3. Survey respondents give equal importance to life saving regardless of the quality of life, provided that the quality of life is above some minimum level.

To reconcile the methods of allocating health-related resources to the attitudes of the public, Nord and several collaborators propose replacing cost-effectiveness analysis with what they call "cost-value" analysis. Rather than assessing policies by the contributions they make to the individual value of health, they propose to assess policies by the contributions they make to the "social value" of health, where that social value is sensitive to ethical considerations that ground objections to cost-effectiveness analysis. For example, the social value of a health improvement is sensitive to the severity of health states. I have myself proposed that cost effectiveness should focus on what I call the "public" rather than the private value of health (Hausman 2015), but in this essay, I shall focus on Nord's notion of social value.

Cost-value analysis differs in two ways from cost-effectiveness analysis. First, it counts saving a life as of equal value regardless of the quality of life, provided that the quality of life is above some minimum level. Second, although it retains the view that effectiveness is a matter of health improvement, cost-value analysis measures improvement in terms of the social value of health, which depends on distributive considerations in addition to the value of health to individuals. The social values of health improvements can be determined by measuring the priority that individuals place on treating those who are more severely ill and by measuring how inclined the population is to enhance the claims to treatment of those who have a lesser ability to benefit from treatment. Thus, Nord et al. (1999, 32) propose a multiplicative model:

$$SV(p) = dU(p) \text{ x } SW(p) \text{ x } PW(p),$$

where *SV* is the social value of a health care intervention *p*, *dU* is the individual value of the health improvement *p*, *SW* is a severity weight, and *PW* is a weight concerned with the potential for health.

A more direct way of measuring the social value of health, which Nord favors, is to survey the relevant population concerning how they would make so-called "person trade-offs." Health economists can ask survey questions such as the following:

> Program P can extend the lives of 100 healthy people for one year. Program Q, at the same cost, alleviates the symptoms of diabetes in N people for a year. For what value of N should society be indifferent between programs P and Q?

If one stipulates that the social value of providing someone with an additional year of life is 1 and surveys reveal that the population is indifferent between P and Q when N = 120, then 120 x *SV*(Diabetes) = 100, which implies that the social value of diabetes is about .83. Since person trade-offs reflect both how people, as individuals, value health states and their views about fairness, they are, Nord argues, a promising method for measuring social value directly.

Table 11.1 Social Values for Different Severities of Illness. *Source*: Nord, Erik. 1999. *Cost-Value Analysis in Health Care: Making Sense Out of QALYs*. Cambridge: Cambridge University Press.

Severity level	*Social value*
Healthy	1.0
Slight Problem	.9999
Moderate Problem	.99
Considerable Problem	.92
Severe Problem	.80
Very Severe Problem	.65
Completely Disabled	.40
Dead	0

Nord argues that social values, as compared to individual values, show a strong upper-end compression. For example, Nord (1999, 119) suggests that the social values for different severities of illnesses might look roughly like those shown in Table 11.1.

If one supposes that the step down from one row to the next is roughly similar in terms of the individual value of the health state, these values reflect the priority given to those who are more severely ill: a one-row improvement in the health of someone with a considerable problem provides a social value of .07, while a one-row improvement in the health of someone with a very severe problem provides a social value of .15.

This upper-end compression captures not only the social concern for severity, but it also mitigates the discrimination that is implicit in cost effectiveness because those who are disabled have a lesser capacity to benefit from treatments of unrelated diseases. Suppose that both Jill and Jack have a very severe problem, which can be completely cured. Owing to a prior unrelated disability, curing the problem leaves Jill with a moderate problem. Curing her provides a social value of .34, while curing Jack, who has no unrelated health problems, provides a social value of .35. Although there is still a greater social value in treating the individual without any disabilities, it is very small and easily outweighed by other factors.

Although Nord's proposal does not at one fell swoop resolve all the ethical issues concerning the application of cost-effectiveness analysis, it would, if successful, make progress in resolving them. Moreover, parts of his proposal appear to be straightforward to implement. Person trade-offs are demanding, but it appears that people can make them. Substituting social values for individual values in cost-effectiveness analysis is unproblematic. Exactly how to incorporate the additional proposal that preventing a death has the same value regardless of an individual's health state is less clear. If an intervention both saves lives and cures some independent disability, is its value just the same as an intervention that saves a life and leaves the disability unchanged?

I shall assume that these technical issues can be resolved and turn to the main questions that I want to raise, which concern the normative foundations of Nord's proposal. Popular ethical attitudes are not, of course, automatically correct. People may be confused or biased. After deeper reflection, they may reject the answers they previously gave to survey questions. Adjusting the values used in cost-value analysis to reflect popular attitudes is no improvement if those values are not themselves defensible. So I propose to look "behind" popular attitudes to the reasons that justify or challenge those attitudes. Although assessing proposals for allocating health-related resources by considering whether they conform to popular attitudes is a sensible shortcut for social scientists, it leaves unanswered the philosophical questions concerning how health interventions ought to be evaluated.

One way to make clear why this philosophical inquiry is necessary is to consider the following proposal. Instead of posing person trade-off questions to survey respondents or gathering attitudes concerning the priority that should be given to those whose health problems are more severe, why not simply survey members of the population about how they would like the health budget to be allocated? There are, to be sure, a variety of practical objections to this proposal. One might, for example, object that people do not know enough to respond to such questions. But similar objections can be raised to the survey questions that are put to people. The main problem is that surveying the population for answers to the question of how to allocate the health budget obviously passes the buck. If it satisfies the relevant ethical constraints, health policy should aim to allocate the health budget in a way that is in accord with the goals that justify the existence of a health budget. There is no way to know whether the way any individual in the population or members of the population on average would allocate the health budget is morally permissible, without determining what the health budget is supposed to accomplish and what ethical constraints there are on its allocation. In my view, much of the work on assigning values to health states is subject to this objection. In surveying members of the target population, health economists are implicitly supposing that members of the population are able to correctly answer the evaluative questions that apparently stump the health economist.

FAIRNESS, SEVERITY, AND DISCRIMINATION

Nord presents a good deal of evidence from studies he and others have carried out, which shows that if faced with a choice of equally expensive treatments for a more severe illness in A or a less severe illness in B, most individuals tend to favor treating A even when the improvement in A's health is somewhat less than the improvement in B's health. Nord and others regard this concern with severity as reflecting a commitment to fairness. Is it? Is it unfair to provide an equally large or slightly larger health benefit to B rather than to A when B is less severely ill than A?

To address this question requires saying something about fairness, and there is, unfortunately, no good general theory of fairness. However, there are some things that can be said. One aspect of fairness is procedural. If in the distribution of benefits and burdens the government fails to abide by its stated procedures or if it fails to be impartial in the administration of its laws, then it treats individuals unfairly even if the outcome might have been reached by a fair procedure. For example, if the criteria determining eligibility for a kidney transplant rank P's and Q's claims on a kidney from a dead donor equally, then Q is treated unfairly if the transplant center is influenced by a large contribution from P's family to give P the kidney. This is true even though if the allocation had been decided by a coin flip, P might have wound up with the kidney anyway. The unfairness lies not in the outcome but rather in the procedure that led to that outcome. This procedural aspect of fairness is irrelevant to the concern with severity, and there is nothing in cost-value analysis that addresses complaints that allocating health-related resources via cost effectiveness involves procedural unfairness rather than unfairness in outcomes (Klondinski 2014).

The objection to cost effectiveness that is grounded in a concern about severity is an objection to the outcome that cost effectiveness endorses, not to its procedures. Even without any general theory of fairness, one might feel that not favoring the treatment of those who are more severely ill is unfair because fairness requires placing an extra weight on the claims of those who are badly off. Although not stated in exactly these terms, this seems to be the case in Derek Parfit's influential defense of prioritarianism (1991). Matthew Adler, in his recent *Well-Being and Fair Distribution* (2012), argues at length that prioritizing the well-being of those who are badly off is the correct way to capture concerns that distribution be fair. If the severity objection reflects a general commitment to a "prioritarian" view of social justice, then it would appear to be a justified complaint of unfairness. For example, Brock (2012, 155; see also 2007, 142) writes,

> Perhaps the most common feature of different theories of justice and of the thinking of ordinary persons about justice is a special concern for the worst-off members of society. . . . Concern for the worse off has a long tradition in political philosophy as well and in more recent decades has been a central focus of the work of John Rawls and many others he has influenced.

I am, however, doubtful about whether prioritarian commitments explain or justify the severity objection because I question both whether prioritarianism supports the severity objection to relying on cost-effectiveness information and whether prioritarianism is itself justified.[5] As several authors point out (Brock 2002; Cookson and Dolan 2000; Kamm 2002; Nord 2013; and Scanlon 2003), a "special concern and priority for the worst off" does not justify singling out for special concern those whose health state is at the moment especially bad. Consider the following four people:

- Helen is sixty years old. She has been indigent her whole life. She is uneducated and is regularly beaten by her husband. Seven of her eight children have died. She is anemic and malnourished but able to carry out her daily activities.
- George is twenty-five and has had cerebral palsy since birth. He can move around only with the help of a special motorized wheelchair. He now has a mild case of pneumonia that will probably clear up on its own but could threaten his life.
- Eileen is seventy-five. She has never been sick until now. She has a bacterial infection that if untreated will leave her deaf and blind.
- Jason is fifty and has been healthy so far. He now has a fungal disease that if not treated immediately will lead over the next twenty years to increasing disability, cognitive limitation, and agonizing pain followed by death.

Who is worst off? Depending on how one understands "worst off," it could be any one of the four. In terms of overall well-being, malnourished Helen seems to be the worst off. In terms of health now, Eileen, who is at imminent risk of deafness and blindness, is the worst off, and as severity is understood by Nord and others in the literature, her health state is the most severe. In terms of future health prospects, Jason, who faces the degenerating disease, is worst off. But since he does not face an immediate serious risk of death or disability, his health condition is less severe than Eileen's. In terms of overall health up to the present, the cerebral palsy victim, George, is the worst off.

Once one recognizes the different ways in which individuals may be advantaged or disadvantaged, it is hard to see how prioritarianism could support favoring prioritizing the treatment of those whose health now is especially bad. Prioritarianism calls for prioritizing benefits to those who are worse off *all things considered* rather than prioritizing treatment of those who are more severely ill now.

The popular attitudes in favor of prioritizing the treatment of those who are severely ill and the intuitive appeal of doing so remain. Where do they come from, and don't they show us that the allocation of health-related resources ought to prioritize the treatment of the severely ill? I think that the sentiments supporting the severity objection have three sources. First, I suspect that the intuitions in support of the severity objection derive in part from the mistaken assumption that in treating Jill, who is more severely ill than Jack, one is providing Jill with a larger health benefit than one would be providing to Jack. This confusion can be only part of the story because those eliciting people's attitudes have worked hard to try to get people not to make this tempting mistake. Second, in those cases in which Jill, whose health is worse, is in desperately bad health, the findings of popular attitudes may reflect a compassionate concern for immediate suffering or risk of death. When an individual reacts this way to the travails of another person, we honor him or her, but whether such reactions should govern state policy is questionable because they ignore opportunity costs, which include other deaths and other suffering. If someone drops everything to come to the rescue of someone else in dire straits, he or she of course incurs costs. But those costs are unlikely to prevent him or her or others from rescuing other people in need. If, in contrast, the health budget is shifted to give priority to those who are currently severely ill, as cost-value analysis requires, there will be fewer resources devoted to prevention and the treatment of lesser risks, and more people will suffer and die.

A third and more promising argument abandons the purported connection to fairness and justifies prioritizing the treatment of those who are seriously ill on the grounds of *solidarity*. What does it say about the connections among citizens that those who are healthy are able to turn their backs on those who have terrible health to improve overall population health? Doesn't the disjunction between the insistent charitable demands on individuals and the indifference of public policies undermine the sentimental ties among citizens that hold a state together? Perhaps. But allocating health resources by their cost effectiveness reflects an impartial concern for everyone's health. Even if less immediately emotionally compelling, isn't cost effectiveness a rational expression of solidarity?

In a recent essay (2016), Nord makes the following argument in defense of the claim that compassion for those who are severely ill implies that it would be unfair not to place a greater weight on their claim to treatment. I have paraphrased, condensed, and tightened Nord's argument without, I hope, distorting it:[6]

1. Assume the public feels greater compassion for the needs of a patient group A than those of group B and feels a greater moral obligation to meet the needs of group A than those of group B.
2. This feeling of a greater moral obligation implies a feeling that the needs of group A have a stronger moral claim to be met than those of group B.
3. Fairness requires greater attention to stronger moral claims.
4. Fairness requires greater attention to the needs of patient group A than to the needs of group B.

I am dubious about both premise 1 and premise 2. My concern about premise 1 is that it does not distinguish the obligation of individuals from the obligation of the state. It is questionable whether the personal concern with the needs of group A implies a public obligation once one recognizes the costs in later suffering and death of prioritizing those needs. With respect to premise 2, not all moral obligations derive from claims, and it is particularly questionable whether the imperfect duties that derive from compassion imply the existence of claims on behalf of those who are the object of one's compassion.

I may have missed something, but the tentative conclusion I draw is that the preference people express for prioritizing the treatment of those who are seriously ill is unjustified and if it is politically feasible, these preferences should not influence state health policy (although they may, of course, influence the private choices of individuals).

CONCLUSIONS

It is thus questionable whether Nord's cost-value analysis improves upon cost-effectiveness analysis. As already noted, it does not address the purported procedural unfairness of relying on cost-effectiveness analysis, which is a large part of the fair chances objection, and its accommodation of the severity objection is unjustified. There remains, however, the discrimination objection. Nord's proposals provide some serious help with the intractable problems in avoiding discrimination without abandoning efficiency, and that is perhaps a sufficient defense. But in some ways, they seem to go too far, while in other ways they are insufficient. Even though the difference between the social value of treating someone with a disability and treating someone who is not disabled is a great deal less than the difference in QALYs, there is still a difference, and that difference is arguably discriminatory and disrespectful. On the other hand, counting the saving of a life as equally valuable may go too far. Should we be indifferent between giving a heart transplant to an otherwise healthy forty-year-old or to a bedridden quadriplegic?

I wish I knew what to say about discrimination, but fortunately I am out of space and can withdraw having broached only a few of the many ethical quandaries concerning the use of cost-effectiveness information to allocate health-related resources. I have argued that effectiveness should be measured in terms of health improvements rather than in terms of improvements in well-being, that severity weighting is unjustified, and that Nord's proposals for addressing the objections to cost-effectiveness analysis, while promising in some regards, are questionable in other regards and in need of further ethical scrutiny like that which this essay attempts.

NOTES

1. This chapter builds on arguments in Chapters 6, 13, 15, and 16 of my *Valuing Health: Well-Being, Freedom, and Suffering* (Hausman 2015).

2. According to Chhatwal et al. (2015), the incremental cost effectiveness (that is, Solvadi's cost effectiveness compared to existing therapies) is somewhere between \$9,700/QALY and \$284,300/QALY depending on the patient's prior medical history.

3. See Robbins (1935, Chapter 5), Harsanyi (1977, Chapter 4), Elster and Roemer (1991), Broome (1993), and Hausman (1995).

4. This claim is plausible and defended by authors such as Griffin (1986), Raz (1986), and Scanlon (1998).

5. Even if prioritarianism supported the severity objection to relying on cost effectiveness, it would not *justify* that objection because prioritarianism is, in my view, as much in need of justification as the severity objection it purportedly grounds. The only argument in its favor is an appeal to intuitions in support of its implications. But those implications have other plausible explanations and accordingly count for little.

6. The first five steps in Nord's version read as follows: (1) Assume the public feels compassion for a patient group A. (2) The compassion leads to feelings of moral obligation to help. (3) This feeling of moral obligation to help is equivalent to the public's feeling that group A has a moral claim on them. (4) Fairness is defined as a moral good that is achieved by adequate response to moral claims (e.g., Broome 1999). (4) Nonaccommodation of group A's moral claim, which derives from the public's compassion, is thus unfair.

REFERENCES

Adler, Matthew. 2012. *Well-Being and Fair Distribution*. Oxford: Oxford University Press.

Brock, Dan. 2002. "The Separability of Health and Well-Being." In *Summary Measures of Population Health: Concepts, Ethics, Measurement and Applications*, edited by Christopher J. L. Murray, Joshua A. Salomon, Colin D. Mathers, and Allan D. Lopez, 115–20. Geneva: World Health Organization.

Brock, Dan. 2003. "Ethical Issues in the Use of Cost-Effectiveness Analysis for the Prioritization of Health Care Resources." In *WHO Guide to Cost-Effectiveness Analysis*, edited by T. Tan-Torres Edejer, R. Baltussen, T. Adam, R. Hutubessy, A. Acharya, D. B. Evans, and C. J. L. Murray, 289–311. Geneva: World Health Organization.

Brock, Dan. 2007. "Health Care Resource Prioritization and Rationing: Why Is It So Difficult?" *Social Research* 74: 125–47.

Brock, Dan. 2012. "Priority to the Worse Off in Health-Care Resource Prioritization." In *Medicine and Social Justice: Essays on the Distribution of Health Care*, edited by Rosamond Rhodes, Margaret P. Battin, and Anita Silvers, 362–72. New York: Oxford University Press.

Broome, John. 1993. "A Cause of Preference Is Not an Object of Preference." *Social Choice and Welfare* 10: 57–68.

Broome, John. 1999. *Ethics Out of Economics*. Cambridge: Cambridge University Press.

Broome, John. 2002. "Measuring the Burden of Disease by Aggregating Well-Being." In *Summary Measures of Population Health: Concepts, Ethics, Measurement and Applications*, edited by Christopher J. L. Murray, Joshua A. Salomon, Colin D. Mathers, and Alan D. Lopez, 91–113. Geneva: World Health Organization.

Chhatwal, Jagpreet, Fasiha Kanwal, Mark Roberts, and Michael Dunn. 2015. "Cost-Effectiveness and Budget Impact of Hepatitis C Virus Treatment with Sofosbuvir and Ledipasvir in the United States." *Annals of Internal Medicine* 162: 397–406.

Cookson, Richard, and Paul Dolan. 2000. "Principles of Justice in Health Care Rationing." *Journal of Medical Ethics* 26: 323–29.

Elster, Jon, and John Roemer. 1991. *Interpersonal Comparisons of Well-Being*. Cambridge: Cambridge University Press.

Griffin, James. 1986. *Well-Being: Its Meaning, Measurement and Moral Importance*. Oxford: Clarendon Press.

Harsanyi, John. 1977. *Rational Behavior and Bargaining Equilibrium in Games and Social Situations*. Cambridge: Cambridge University Press.

Hausman, Daniel. 1995. "The Impossibility of Interpersonal Utility Comparisons." *Mind* 104: 473–90.

Hausman, Daniel. 2015. *Valuing Health: Well-Being, Freedom, and Suffering*. New York: Oxford University Press.

Kamm, Frances. 2002. "Health and Equity." In *Summary Measures of Population Health: Concepts, Ethics, Measurement and Applications*, edited by Christopher J. L. Murray, Joshua A. Salomon, Colin D. Mathers, and Allan D. Lopez, 685–706. Geneva: World Health Organization.

Klondinski, Andrea. 2014. "Economic Imperialism in Health Care Resource Allocation: How Can Equity Considerations Be Incorporated into Economic Evaluation?" *Journal of Economic Methodology* 21: 158–74.

Nord, Erik. 1999. *Cost-Value Analysis in Health Care: Making Sense Out of QALYs*. Cambridge: Cambridge University Press.

Nord, Erik. 2013. "Priority to the Worse Off: Severity of Current and Future Illness versus Shortfall in Lifetime Health." In *Inequalities in Health: Concepts, Measures, and Ethics*, edited by Nir Eyal, Samia A. Hurst, Ole F. Norheim, and Dan Winkler, 66–73. Oxford: Oxford University Press.

Nord, Erik. 2016. "Public Values for Health States versus Societal Valuations of Health Improvements: A Critique of Dan Hausman's 'Valuing Health.'" *Public Health Ethics*, 1–10. doi:10.1093/phe/phw008.

Nord, Erik, J. Pinto, Jeff Richardson, Paul Menzel, and Peter Ubel. 1999. "Incorporating Societal Concerns for Fairness in Numerical Evaluations of Health Programs." *Health Economics* 8: 25–39.

Parfit, Derek. 1991. *Equality or Priority*. The Lindley Lecture. Department of Philosophy: University of Kansas.
Raz, Joseph. 1986. *The Morality of Freedom*. New York: Oxford University Press.
Robbins, Lionel. 1935. *An Essay on the Nature and Significance of Economic Science*. 2nd edition. London: Macmillan.
Scanlon, Thomas. 1998. *What We Owe to Each Other*. Cambridge, MA: Harvard University Press.
Scanlon, Thomas. 2003. *The Difficulty of Tolerance: Essays in Political Philosophy*. Cambridge: Cambridge University Press.

Chapter 12

The Value of Statistical Lives and the Valuing of Life

Michael Dickson

There are at least four (sometimes entwined) social contexts in which some form of material value appears to be put on human life itself. One is actuarial, for determining insurance payouts. A second is regulatory, for establishing appropriate deterrence to causing loss of life or threats to life. A third is judicial, for determining compensation for the loss of life. A fourth is budgetary, for determining how much society will spend to save or extend lives (i.e., to reduce either immediate or long-term risk of death).

The first three measures of the "value" of a life are not, in the end, directly indicative of what, if any, the material value of a life per se is or should be in society. Regarding the actuarial (Viscusi 2005, 2), in contrast to property damage, where insurance payouts are typically able to restore lost utility and thus may be a measure of what was lost, "fatalities . . . affect one's utility function, decreasing both the level of utility and the marginal utility for any given level of income." Therefore, insurance payouts on death ought not be viewed as a straightforward compensation for what was lost, and indeed, because of the decrease in utility, actuarial valuations of life would be expected to be relatively modest compared with any actual "value" of the life that was lost and should not be taken as a measure of that value.

The deterrent measure is, similarly, based not on any supposed *actual* value of life but rather on the potential degree of temptation (or ignorance) that might lead to creating life-threatening risks (or, more directly, the taking of life), insofar as the value assigned a life for deterrence is intended to prevent the creation of such risks. Typically (for example, in the case of fines for littering), the value required for deterrence is arguably much *greater* than the potential harm and in any case, is not directly and solely based on the magnitude of the harm, and so ought not be taken as a (direct) measure of that magnitude.[1]

The judicial measure also does not pretend to provide a value for life itself. Judicially imposed payments for loss of life are based on either violation of regulatory laws (such as laws about workers' safety or more commonly in recent decades, environmental protection) or the outcome of a civil wrongful death lawsuit. The former type of payment is a (failed) deterrent, while the latter type reflects the value of the deceased specifically to individuals who were close to the deceased in some way (normally near kin and legal spouses).[2]

The budgetary measure, in contrast to the first three, is at least aimed in the right direction because it involves some sort of direct trade-off between saving lives and spending money. Moreover, this trade-off is typically evaluated without any knowledge of *whose* life is at stake and thus more plausibly concerns the value of life per se rather than the value of *this* or *that* life; the trade-off is also typically made by society (or at least large groups of individuals) and thus more plausibly concerns the value of life *in general* and not the value of some specific life *for this* or *that* individual. The budgetary measure is therefore plausibly relevant to the project of attaching some monetary value to human life per se.

Over the past several decades, the budgetary valuation has been framed in terms of the "value of a statistical life," a term whose meaning has become somewhat standardized in ways described below. The point of the term is that for practical reasons, we are often faced with decisions about spending money not for the saving of this or that life but rather for the sake of small decreases or increases in the risk of death. Spending is directed to these small decreases and increases rather than to "whole lives." Nonetheless, these small decreases and increases do have consequences, at least *statistically*, for actual lives saved and lost.

In many cases, the numbers obtained as the "value of a statistical life" (VSL) are used to determine whether to pursue potentially lifesaving (or avoid potentially life-threatening) policies or actions. Governmental agencies responsible for environmental policy, food and drug safety, transportation safety, health policy, and more, explicitly incorporate VSL as part of an overall budgetary calculus to inform decisions about policies and activities or to analyze those decisions post hoc.[3] In addition, in conjunction with measures of quality of life, VSL is transformed into the value of a Quality-Adjusted Life Year and thereby informs some theory and decision making (or post hoc analysis) in medical contexts.

However, these practices—the modeling, measurement, and use of VSL—are problematic in several ways. The main section of this paper concerns some of those problems, focusing on the closely connected issues of modeling and measurement. The central claim there is that the currently most popular model of VSL and the demands that it places on measurement largely nullify the advantages of VSL, mentioned above, as a measure of the value of life per se. The subsequent section suggests that these problems might stem from a case of mistaken identity: the magnitude that we should be trying to model and measure is not the *value* of a *life* but rather the *value of valuing* life. For example, when society chooses[4] to spend $10 million to reduce the risk of highway deaths, the suggestion is that this value is (or ought to be) attached not to the lives saved but rather proximately, to the act itself, "taking steps to reduce highway deaths" and ultimately to the valuing of life that the act exemplifies. This alternative way of conceiving of what we are doing when we attempt to make budgetary decisions about lifesaving expenditures would affect current practices in the measurement of VSL, or rather, VVL ("value of valuing life").

THE VALUE OF STATISTICAL LIVES

While the practice of putting a monetary value on life remains highly controversial, there is widespread agreement that the best model to use is "willingness to pay" (WTP). This model has largely superseded an earlier one, the "human capital" (HC) model, in which value was determined by lost productivity[5] (though HC remains an important component of the judicial measure). Even when "productivity" is defined very broadly so that, for example, it might capture the lost value of child rearing or of leisure time, economists seem largely to agree that the HC model is unfixable, at least for determining an actual value for life per se (which is distinct from its use in judicial measures). The WTP model gets "more directly" at the monetary value that individuals implicitly put on life by examining what people are willing to pay to gain some reduction in risk of death or what they are willing to accept as compensation for taking on some additional risk of death.

Economists and others are correct for several reasons to favor the WTP model over the HC model for the budgetary measure of the value of life. First, it avoids the need to ask the question (directly) "What is the value of a (whole) human life?"—a question that is ethically problematic and to which usable answers are unlikely to be given. Second, unlike the HC model, it focuses on a plausibly relevant target for societal policy making, not the value *lost to close associates* because of a loss of life, but the value of the life itself. Third, it places the valuation into the hands of individuals. A "correct" HC valuation of a life requires economic expertise and is thus outside of the purview of most individuals; the HC valuation is thus less likely to reflect the values of society and more likely to reflect the considerations internal to a small and not necessarily representative subset of society.

However, despite this more promising start, the WTP model faces serious challenges regarding both its formulation and the closely related matter of how it is to be measured.

Formulation of the WTP Model of VSL

A standard WTP model of VSL is based on the idea that individuals maximize their expected utility, $Z(p,w)$, where p is the probability of survival and w represents wealth. In the simple case where p is evaluated over a single period,[6]

$$Z(p, w) =_{\mathrm{df}} pU(w) + (1-p)V(w), \quad (1)$$

where $U(w)$ is the utility of w conditioned on survival, and $V(w)$ is the utility of w conditioned on death (by the end of the period).[7] It is standardly assumed that $U()$ and $V()$ are twice differentiable and that for all w:[8]

$$U(w) > V(w) \text{ and } \partial Y/\partial w > \partial \zeta/\partial w \geq 0. \quad (2)$$

In other words, for given wealth, individuals strictly prefer life to death and value marginal increases in wealth nonnegatively and more highly in life than in death.

The feature of this model that helps with measuring VSL is that constant values of Z define "indifference curves" (in the "p-w plane"), along which individuals are indifferent between the various combinations of p and w that lie along the curve. Therefore, we can examine what "trade-offs" an individual is willing to make between p and w to stay on an indifference curve and thereby derive some accounting of what the individual would pay (in w) for increases in p and how much the individual would demand for decreases in p.

Accordingly, VSL is defined in this model by the ratio of the partial derivative of $Z(p, w)$ with respect to p ("how much expected utility changes with a change in p") to the partial derivative of $Z(p, w)$ with respect to w ("how much expected utility changes with a change in w"). The result is a ratio of marginal change in wealth to marginal change in probability of survival (i.e., the marginal rate of substitution of w for p). The result is

$$\mathrm{VSL}(p, w) =_{\mathrm{df}} \frac{U(w) - V(w)}{pU'(w) + (1-p)V'(w)'} \quad (3)$$

where the derivatives $U'(w)$ and $V'(w)$ are taken with respect to w. In other words, VSL at (p, w) is the difference in utility of wealth divided by the expected marginal utility.

This simple form of the model can be extended and elaborated in many ways, for example, to cover multiple time periods or to add other elements of "realism" to the model.[9] However, for present purposes, this model is enough. The features considered here are shared by other versions.

This model predicts some features of empirical data that are often taken to supply information about VSL. It predicts the "dead-anyway" effect,[10] according to which willingness to spend on reducing risk will be high if p is very small, and the "wealth effect,"[11] according to which willingness to spend on reducing risk increases with w. For example, the latter is said to follow from (3) because the difference between $U(w)$ and $V(w)$ is greater with greater w ("the wealthy have more to lose"), whereas the difference between $U'(w)$ and $V'(w)$ is small for large w because of the decreasing marginal utility of wealth (both $U'(w)$ and $V'(w)$ approach zero as w increases, narrowing their difference).

Pratt and Zeckhauser (1996) suggest, with some concern, that the dead-anyway effect could lead to irrational social policy. Problems with the wealth effect (which they also mention) are clear as well because it suggests that two individuals identical in all but wealth could have lives of very different "value." These problems and others arise from at least two fundamental features of this model of VSL.

The first is the model's appeal not to what value individuals put on life per se, but rather to what (marginal) value they put on *their own* lives, *in their present situation* (i.e., given p and w). That is, exchanges of w for p are made *along curves of constant expected utility, Z.* These exchanges can be, and generally will be, sensitive to *which* curve of constant Z an individual occupies. To get from a collection of such values to an overall "societal" value for life requires somehow aggregating the individual values, but it is not at all clear how to do so.[12]

This challenge might appear to be just the familiar one of how to aggregate individual preferences into a "group" preference, and as such, there are familiar solutions and debates about those solutions.[13] However, those solutions often involve making decisions about the "level of importance" that is attached to various types of individual contribution. (Somewhat simplistically, one might exclude or discount extreme values or use the median rather than the mean.) Therefore, because VSL is, in this model, a kind of assessment of the importance that one attaches to *oneself* (one's own life), the "level of importance" that *analysts* attach to individuals for aggregation is dangerously close to the very same assessment, and thus in the act of aggregation one runs the risk of *usurping* (rather than "merely" discounting or exaggerating) the preferences expressed by some individuals. This danger is inherent in the standard WTP model of VSL, as it depends on the value that individuals place on *their own* lives.

The danger is not merely abstract or remote. In any WTP model (whether for statistical lives or some other good), the resources spent by a wealthier individual are typically lower in value per unit than the resources spent by a less wealthy individual. Aggregation thus *requires* one either to *accept* that there will be a difference in influence on the aggregated value depending on wealth or to adjust how individuals are counted. The point here is not that the judgment could not be made in some principled fashion but rather that it must be made at all and that in making it, one "rewrites" the judgments made by the individuals themselves.

A second issue with the model is that it is dangerously close to enabling what nobody who is involved in this work expressly wishes to do, namely, placing a specific monetary value on specific lives. Researchers are sensitive to this issue:[14] "VSL does not measure what an individual is willing to pay to avoid death with certainty, nor what he is willing to accept [WTA] to face death with certainty. It measures the WTP or the WTA for an infinitesimal change in risk" (Andersson and Treich 2011, 5). However, as much as researchers wish, wisely, to focus on *marginal*, even *infinitesimal*, changes in risk, the definition itself simply *does* permit one to put a monetary value on life. Nothing apart from moral discomfort and the suspicion that something is going wrong prevents one from integrating the infinitesimals of the definition.[15] If the model accurately captures its target (i.e., the infinitesimal marginal value that people place on infinitesimal changes in probability of survival), then the monetary value of avoiding outright death simply *is* well defined—infinitesimals render their mathematical integration meaningful. One might conclude, not unreasonably, that any moral problems with this concept therefore reflect on the original model.

Measuring VSL

When it comes to measuring VSL there are two related challenges that should give one pause. The first arises from the fact that nobody *ever* buys (or forgoes buying) marginal increases or decreases in the probability of survival, which is not a commodity (and any way of making it a marketed commodity would be morally repugnant). The analyst therefore must find a proxy for it. The second challenge follows: because analysts are examining a *proxy* for the activity of real interest, there is always a risk that individuals consider and approach these proxies under conceptions different from what the analyst assumes, with motivations that are unrelated (or only partially relevant) to survival.

Proxies come in two forms: revealed preference and stated preference. Lacking space for a comprehensive review,[16] the following is a brief consideration of the two challenges just mentioned in each case, beginning with revealed preferences.

Several consumer choices are thought to reveal "underlying" VSL, from choosing to use safety devices, such as bicycle helmets or seat belts, to choosing safer or less safe cars, to choosing risky jobs, and more. In every case, however, there is a looming problem: it is far from clear that consumers who engage in these decisions are in fact driven solely, or even primarily, by a consideration of the risk of death.[17]

This problem always arises from "revealed preferences," whether one takes the (arguably orthodox) view that preferences in economics are *defined* in terms of observable choices made by consumers or the alternative (arguably "common sense") view that preferences are connected to consumers' choices by their beliefs so that preferences are revealed by choice only considering belief.[18]

In the latter case, a problem with using consumers' choices to discover preferences regarding risk of death is that consumers are notorious for *not* approaching risk in a manner that is consistent with the probability theory that is used to represent it in the WTP model. From the early study of Knight (1921), to the classic research program of Kahneman and Tversky (1982), to the latest research on consumers, there is overwhelming evidence that the way consumers conceive of risk and the way in which it is formalized in many economic models are radically different. For example, Kahn and Kupor (2017) find that "a medical drug with a potential side effect of seizures is viewed [by consumers] as less threatening when it also has smaller potential side effects, such as congestion and fatigue," an attitude that flatly contradicts standard probability theory. In short, once one takes consumers' *beliefs* into account, there is, *at best*, no clear relationship between consumers' *actual* choices on the one hand and the quantity (VSL) that we hope to measure on the other given that our model for VSL *essentially* involves "risk of death" formulated in terms of probability. Consumers' choices do not reveal their preferences regarding risk of death, formulated probabilistically.

On the former "orthodox" view of preferences, this form of the problem does not arise because consumers' beliefs about risk and probability are not relevant to the analysis. The idea is that consumers *are effectively* making choices that involve trading wealth for risk of death and those choices are taken to *define* "consumer preferences." Thus, *by definition*, consumers have "preferences" regarding trade-offs between wealth and risk of death, and those preferences are, by definition, revealed in their choices.

However, this approach robs the WTP model of its purported advantages, which, recall, were that it focuses on the correct *target* (the value of life per se) and that it appeals to the (aggregated) preferences of individuals rather than expert judgment. Once one removes agents' *actual* beliefs from the picture and construes their activity as spending on reducing probability of death even if they do not conceive it in that way, these advantages disintegrate, for there are numerous descriptions of consumer activity that are compatible with observation. One could just as well describe it as spending on complying with regulations, or mollifying the worries of a nervous spouse, or not being viewed as careless. The analysis has thus lost its focus on the value of life per se, and preferences have once again been taken out of the hands of individuals and placed into the hands of the expert, who chooses to describe consumer activity in one way rather than another.

One might reasonably suppose that this problem can be avoided if we consider, instead, *stated* preferences, which are elicited using methods such as "contingent valuation," in which respondents either state a maximum value they would pay or demand for a given change in risk, or make a (hypothetical) choice between two options (e.g., "pay $100 or live with a faulty traffic light?"). The hope is that by using carefully crafted questions, the analyst can control what (hypothetical) choice is being made by the respondent and thus reducing uncertainty about how that choice is understood.

Nonetheless, the question that the analyst *thinks* is being asked need not be the question that is understood by the respondent, whose conception of the question and choice to answer in one way rather than another may be significantly affected by "social desirability bias,"[19] a natural and pervasive desire to prefer social acceptability over honesty and to perceive questions from others as probes of social acceptability rather than solicitations of genuine belief. A pollster may ask, for example, "Would you be willing to pay $1,000 to repair a faulty circuit box?" but the respondent hears "Are you such a skinflint that you would risk your family's life to save $1,000?"

A related problem is that stated preferences regarding risk of death are (in any ethically reasonable analysis) hypothetical, raising the issue of "hypothetical bias,"[20] the difference between *stated* willingness to pay and actual willingness to pay as exhibited in real consumer behavior. It is a well-established phenomenon, which is typically interpreted as a tendency for respondents to overstate (though in some circumstances, understate) the amount they are willing to pay for a given good.

There is widespread interest in mitigating the effects of hypothetical bias,[21] using both *ex ante* methods, typically involving redesigning the survey or priming respondents in some manner to be honest in their appraisal, as well as *ex post* methods, typically involving statistical manipulations or recoding answers based on postsurvey questions about the respondents' level of certainty about their answers.

Some of these methods have been found to be somewhat effective, but their implementation and effectiveness have also been found to be strongly tied to specific contexts, and herein lies their downfall in the present case: methods for mitigating hypothetical bias are plausibly verified only in circumstances where data are *also* available for actual consumer behavior.[22] However, as noted earlier, "reduction in the risk of death" is not a market commodity, and thus we have no unproblematic real-world data about purchasing with which to compare the results of our surveys.

These problems loom large for the WTP model of VSL because hypothetical bias is likely to be caused or exacerbated by the involvements of two concepts about which people are notoriously poor at reasoning: death and probability. Even from a purely economic standpoint, it is difficult for agents to reason well about death because death "changes one's utility function," as noted above, and it is difficult if not impossible to understand that change in detail. Also, as noted above, there is ample evidence from the psychological literature that agents do not reason particularly well about probability.

THE VALUING OF LIFE

No claim is being made here that the problems sketched above are insoluble. At the same time, while these problems may flow from familiar challenges to measuring WTP for *any* nonmarket good, whether via revealed preferences or stated preferences, the discussion of the previous section highlighted several reasons that the problems are particularly recalcitrant in the case of the WTP model of VSL. That discussion also directly raised some questions about the definition of the model itself.

Perhaps a root cause of the problems lies in the choice of what to measure in the first place. After all, the notion of "buying probability for personal survival" *inevitably* leads to the problematic well definedness of "the cost of certain death." Similarly, the individualistic WTP model of VSL necessitates aggregation, which as we have seen, is problematic; the definition in terms of *inevitably* "inaccessible" economic transactions ("buying probabilities") necessitates the problematic use of proxies; and finally, the product itself ("reduction in the probability of death") involves the problematic concepts of probability and death.

Perhaps a better concept to consider is the "value of the valuing of life" (VVL). Rather than considering how much agents are willing to pay "per unit" of reduction in the probability of death, one might consider how much agents are willing to pay for a given "amount" of "valuing of life."

Setting aside (for the moment) the issue of how to measure "amounts" of "valuing of life," VVL departs from VSL in a few potentially helpful ways. For example, while it still refers to the value that *individuals* place on the good in question ("valuing life"), the good itself no longer *directly* concerns the individual—individuals are not placing a value on their own lives—and thus the particularly troublesome form of the problem of aggregation does not arise; neither do difficulties in conceiving the economic value of one's own life. In addition, VVL does not directly involve probability of death and therefore avoids at least some misunderstandings about probability, as well as "integrating to find the value of a whole life."

No doubt there are, nonetheless, many challenges faced by measuring VVL. One apparent challenge is that unlike probability, which by its nature *is* a measure, there is no obvious measure of the "amount" of valuing of life—"valuing of life" is not a good that comes with a clear quantitative measure. Therefore, while clearly, "valuing life" does admit of degrees—some people or institutions do it more than others—it could prove a challenge to formulate some measure not only of how much we "value valuing" but also how much "valuing" is being done in the first place.

But there is no reason to give up on VVL for this reason—it is, in the end, an empirical question what "markers" individuals use for assessing how much they, or others, or institutions value life. (Some of those markers would likely be economic—for example, giving up a certain percentage of profits for the sake of safety. Others would not be obviously or directly economic—for example, devoting time in an annual meeting to considering product safety or reacting with concern versus disdain for complaints about danger.) The present proposal thus carries two connected, empirical problems: (1) How do individuals assess "the amount" that another individual or institution values life? (2) How much value do individuals put on various "amounts" of valuing life?

Note that there are two types of measurement in play here—measurement of "amounts" of valuing life and measurement of how much value is placed on specified "amounts" of valuing life. These two types of measurement are, roughly, playing the roles that probability of survival and the value placed on probability of survival play in VSL.

Regarding (1), it could turn out that there is no common unit of measure of the "amount" of valuing life and thus no universal measure of the value of valuing life. That finding would not derail the present proposal, a fact that marks a radical departure from VSL, which enables one (in principle) to determine a price tag for any activity that changes the probability of death. In contrast, VVL might be heterogeneous—it might depend on the *type* of valuing in question.

We would not thereby be hampered in making the sorts of budgetary decisions for which VSL was intended in the first place. Those decisions would simply be made directly in terms of the monetary value that is attached (by society) to various types of valuing of life rather than in terms of a universal measure. In the end, the "common measure" would be currency rather than a (imagined? manufactured?) "value per unit of valuing."

Of course, there is still the issue of how one could measure VVL even if measurement must become more context specific. While there is no theory on offer here, the concept of "valuing life" is clear enough in practice that measuring its actual or proper value to society is not hopeless, or at least no *more* hopeless than any other endeavor that somehow mixes sociology, economics, and moral philosophy. One could look, for example, at informed investors' willingness to give up some profit for the sake of investing in companies they perceive as taking certain kinds of steps to save lives. One could look at society's willingness to spend on lifesaving services (fire and rescue, for example). One could look at individuals' willingness to give up some of their own wealth to support agencies (such as the Red Cross) that are generally perceived to place a high priority on valuing lives. There are difficulties in each case, but the suggestion here is that some particularly trenchant difficulties with measuring VSL (and some troubling consequences of the definition) might be avoided.

No doubt econometric challenges loom in each of these cases, others like them, and others perhaps more radical, perhaps even departing from the WTP model that underwrites both current attempts to measure VSL and the suggestions above for measuring VVL. On the other hand, perhaps VVL is nonetheless a step in a better direction, conceptually, metrically, practically, and ethically.

NOTES

1. Magnitude of harm *is* a factor in determining deterrence—penalties for causing death are higher than penalties for littering for a good reason. Nonetheless, the magnitude of the penalty required for deterrence is not simply a compensation for the (potential, or expected value of) loss. Further, no claim is being made here that material deterrence *over*values human life. The case is quite different from littering. The point is simply that material deterrence does not *measure* the value of that for which it is a deterrent.

2. For example (and it is typical), the Revised Ohio Code § 2125.02(A.1) states that damages due to wrongful death are to be for "the exclusive benefit of the surviving spouse, the children, and the parents." Moreover, these damages are limited to compensation for certain types of losses that are specific to the beneficiaries, namely, loss of financial support, loss of services, loss of society (e.g., companionship), loss of prospective inheritance, and mental anguish.

3. For a somewhat critical review of the use of VSL by various regulatory agencies, but sympathetic to the overall project, see Kenkel (2003).

4. The mechanism by which such judgments are made is not at issue here; any case of spending (or avoiding spending) public resources will be considered a "societal choice."

5. For a review of general methods and literature (though applied to transport safety), see Andersson and Treich (2011).

6. The model was introduced by Dréze (1962) and has been studied extensively. See Viscusi and Aldy (2003) for a more recent version and discussion.

7. The same mathematical model is used for both death and serious injury. In the latter case (not under consideration here), $V(w)$ should reflect any (potentially wealth-dependent) compensation or benefits received by the individual due to injury. In any case, the conceptual foundations of this model may be very different for cases of (even serious) injury as opposed to cases of death, a point that is sometimes ignored.

8. The model also assumes that second partial derivatives are negative, which reflects the decreasing marginal utility of wealth.

9. See Lee and Hatcher (2001, 121–25) for a general review (applied to WTP for information). See Andersson and Treich (2001, 2.1) and references therein for further discussion, more directly about VSL.

10. See Pratt and Zeckhauser (1996), whose analysis suggests that, the effect arises from (2), for if p is small, then wealth is imminently perishable—it will likely "become useless," so best consume it now even if doing so has minimal benefits.

11. See Viscusi and Aldy (2003). Based on a meta-analysis of studies, they estimate income elasticity of VSL at 0.5 to 0.6.

12. A related problem arises from the need to aggregate the results of different empirical studies in a meta-analysis. See Robinson and Hammit (2015).

13. For a thorough bibliographic review, see La Red Martinez and Acosta (2015).

14. Earlier, expressing some sensitivity to the moral danger of putting prices on lives, Andersson and Treich (2011, 1) write, "The expression 'value of life' is an unfortunate reduced form of the value of a statistical life (VSL), which defines the monetary value of a (small and similar among the population) mortality risk reduction that would prevent one statistical death and, therefore, should not be interpreted as how much individuals are willing to pay to save an identified life."

15. The functions *could* be nonintegrable, but there is no reason to think that they would be.

16. For a review, see Ashenfelter (2006); cf. OECD (2012).

17. This issue, among others, motivates many court rejections of the WTP models of VSL. For example, in rejecting the outcome of a WTP model, the Seventh Circuit expresses concern that "spending on items like air bags and smoke detectors is probably influenced as much by advertising and marketing decisions . . . and by government-mandated safety requirements as it is by any consideration by consumers of how much life is worth" (*Mercado v. Ahmed* 974 F. 2d 863, 7th Cir. 1992). There are numerous other well-known problems with revealed preferences. For example, often, consumers are faced with a very limited range of choices ("wear a seat belt or not"), and it is unclear how to translate such choices into a meaningful utility function over a greater range of states, as is typically required. More generally, the very notion of revealed preferences might also presuppose a controversial model of economic agents, a point not considered further here. (For a recent discussion, see Infante, Lecouteux, and Sugden (2015) and subsequent discussions in the same journal.)

18. Hausman (2012, especially Chapter 3) argues for the latter view. I am sympathetic but need not commit to it here.

19. See Krumpal (2013) for a review.

20. See Loomis (2014) and references therein. Loomis also reviews some attempts to mitigate the problem of hypothetical bias though, as he points out, methods to mitigate it are less plausible in some cases than in others and are most plausible when there is also direct evidence (from a market).

21. See Blumenschein et al. (2008), Hensher (2010), and Loomis (2014).

22. See Blumenschein et al. (2008), who surveyed some subjects while allowing others to make an actual choice about purchasing a specific type of medical care.

REFERENCES

Andersson, Henrik, and Nicolas Treich. 2011. "The Value of a Statistical Life." In *A Handbook of Transport Economics*, edited by André de Palma, Robin Lindsey, Emile Quinet, and Roger Vickerman, 396–424. Cheltenham, UK: Edward Elgar Publishing.

Ashenfelter, Orley. 2006. "Measuring the VSL: Problems and Prospects." *The Economic Journal* 116: C10–C23.

Blumenschein, Karen, Glenn Blomquist, Magnus Johannesson, Nancy Horn, and Patricia Freeman. 2008. "Eliciting Willingness to Pay without Bias." *The Economic Journal* 118: 114–37.

Dréze, Jacques. 1962. "L'Utilité Sociale d'une Vie Humaine." *Revue Française de Recerche Opérationelle* 6: 93–118.

Hausman, Daniel. 2012. *Preference, Value, Choice, and Welfare*. New York: Cambridge University Press.

Hensher, David, H. 2010. "Hypothetical Bias, Choice Experiments, and Willingness to Pay." *Transportation Research Part B: Methodological* 44: 735–52.

Infante, Gerardo, Guilhem Lecouteux, and Robert Sugden. 2015. "Preference Purification and the Inner Rational Agent: A Critique of the Conventional Wisdom of Behavioural Welfare Economics." *Journal of Economic Methodology* 23: 1–25.

Kahn, Uzma, and Daniella M. Kupor. 2017. "Risk (Mis)Perception: When Greater Risk Reduces Risk Valuation." *Journal of Consumer Research* 43: 769–86.

Kahneman, Daniel, and Amos Tversky. 1982. "On the Study of Statistical Intuitions." *Cognition* 11: 123–41.

Kenkel, Don. 2003. "Using Estimates of the Value of a Statistical Life in Evaluating Consumer Policy Regulations." *Journal of Consumer Policy* 26: 1–21.

Knight, Frank. 1921. *Risk, Uncertainty, and Profit.* Boston: Houghton Mifflin Company.

Krumpal, Ivan. 2013. "Determinants of Social Desirability Bias in Sensitive Surveys: A Literature Review." *Quality and Quantity* 47: 2025–47.

La Red Martinez, David L., and Julio Acosta. 2015. "Review of Modeling Preferences for Decision Models." *European Scientific Journal* 11: 1–18.

Lee, Kyung Hee, and Charles Hatcher. 2001. "Willingness to Pay for Information: An Analyst's Guide." *Journal of Consumer Affairs* 35: 120–40.

Loomis, John. 2014. "Strategies for Overcoming Hypothetical Bias in Stated Preference Surveys." *Journal of Agricultural and Resource Economics* 39: 34–46.

OECD. *Mortality Risk Valuation in Environment, Health, and Transport Policies.* 2012. Paris: OECD Publishing.

Ohio Administrative Code. Revised Ohio Code § 2125.02(A.1). Last modified December 19, 2016. http://codes.ohio.gov/orc/2125.02v1.

Pratt, John W., and Richard J. Zeckhauser. 1996. "Willingness to Pay and the Distribution of Risk and Wealth." *Journal of Political Economy* 104: 747–63.

Robinson, Lisa A., and James K. Hammit. 2015. "Research Synthesis and the Value per Statistical Life." *Risk Analysis* 35: 1086–1100.

Viscusi, W. Kip. 2005. "The Value of Life." Discussion Paper No. 517, Harvard Law School. http://www.law.harvard.edu/programs/olin_center/.

Viscusi, W. Kip, and Joseph E. Aldy. 2003. "The Value of a Statistical Life: A Critical Review of Market Estimates throughout the World." *Journal of Risk and Uncertainty* 27: 5–76.

Chapter 13

How Good Decisions Result in Bad Outcomes

Anya Plutynski

The fact that the health care system is "increasingly fractured by specialization, discontinuity of caregivers, and elaborate divisions of labor among health professionals" (Kukla 2007, 31) is not news. The fractured nature of medical care is one central cause of medical error, unnecessary tests and other interventions, and thus inefficient and ineffective care. The Institute of Medicine's (IOM) 2000 publication, *To Err Is Human: Building a Safer Health System*, provided evidence that 3–4 percent of patients hospitalized in the United States were harmed by care received, and 44,000–98,000 patients died because of medical errors. A more recent paper (Makary and Daniel 2016) identifies medical error as the third major cause of death in the United States, after heart disease and cancer.

Such a significant cost to life and health calls for serious reflection. What are the major causes of such errors? How can medical care be more effective? And how do we best prevent error and promote better health? This essay will focus on a major class of the errors discussed in Makary and Daniel (2016); they call this class of error "system level" problems, such as failures in communication due to the distributed nature of decision making in many medical contexts. According to the IOM report, more effective teamwork and better communication between caregivers could have prevented as much as half of the medical errors (Kohn, Corrigan, and Donaldson 2000).

This fracturing of care yields poor overall outcomes for individual patients. More precisely, distributed decision making permits and indeed may well promote systematic forms of error. In this chapter, I discuss one example, where prima facie "good" decisions by each individual decision maker can result in worse outcomes for patients. Hence, the title of the paper, how "good" decisions result in bad outcomes. I suggest that taking such apparently paradoxical cases seriously and putting into practice measures to prevent them will require rethinking the standard ways that effectiveness in medical care is assessed.

Currently, US hospitals are evaluated by both public and private not-for-profit bodies (e.g., Medicare, Medicaid, Agency for Health Care Quality, the Joint Commission, the National Quality Forum) to consider their "effectiveness" at meeting certain end points, or benchmark goals. Typical goals include reducing the frequency of readmission (the return of patients to the hospital for further treatment), providing beta blockers to patients with acute myocardial infarction, providing discharge instructions, or reducing rates of infectious disease. The Affordable Care Act also ties assessments of clinicians' effectiveness to meeting specific goals or end points, for example, providing specific tests or counseling to patients with various risk factors. The central way(s) in which effectiveness is conceptualized focuses on relatively narrow, specific outcomes and individual "one-off" decisions rather than patterns of decision making over time. However, while (on average) the provision of a drug, test, or treatment may improve outcomes in a patient population (e.g., it may resolve symptoms or reduce risk of future disease or mortality), this may not be true for all patients, all the time. Moreover, this benchmark approach fails to attend to the dynamic and distributed aspects of care and takes a particularly narrow view of health. If quality of care is a process, then we need to rethink this means of assessing effective care (McClimans and Browne 2012). Following, I argue that assessing the effectiveness of health care should involve multiple components in a way that acknowledges both the distributed character of health care, the variability of patients, and the dynamic, temporal dimension of care over time.

Distributed decision making in health care has a specific family of problems associated with it. All too often, while each individual decision or intervention in such a distributed process may be "effective," "risk averse," or generally sound from a narrowly conceived approach to measuring utility and rationality, the overall process is not. Understanding that we need to consider the effectiveness of the process, and not only individual outcomes, is essential to both diagnosing error and considering ways to improve measures of outcomes. To be sure, this insight about distributed decision making in medicine is not entirely novel. There are a variety of attempts to address these concerns in the health care policy literature. There has been a gradual shift from standard individual-based models of decision theoretic approaches to analysis of the role of health care teams in medical decision making.

However, such approaches, in my view, too narrowly focus on the ways in which errors can emerge from failure of coordination and communication (Patel et al. 2008). The presupposition seems to be that the difficulty is simply a matter of failure of communication: if we communicated better, we would not make mistakes. This has been accompanied with integration of psychological research on heuristics and biases (Patel, Kaufman, and Arocha 2002). Several models have been developed to represent the aspects of teamwork that influence team effectiveness (Lemieux-Charles and McGuire 2006). These models highlight how the organizational context in which a team operates (e.g., goals, structure, rewards, training environment) indirectly influences its effectiveness. Clear communication, well-defined leadership and decision making roles, psychosocial traits and team composition have an influence on quality of care. Incorporating such measures into assessment of practice, however, has been a challenge. Moreover, they do not address the specific systemic problem I focus on here.

I begin with an illustrative example. The aim is not to show that this particular decision making process is typical, nor that this decision process always results in poor outcomes for patients, but only that there are better ways to achieve effective care.

A child with acute respiratory distress is taken to her pediatrician. The evening before, her mother noticed that the child was coughing and had a cold. Knowing that this child tended to develop asthma under these conditions, she gave the child "emergency medication" (albuterol), which seemed to resolve the cough and reduce the child's distress. She gave the same meds in the morning before school; the child had no fever but seemed mildly congested. At school, the child seemed to be wheezing and was given medication by the school nurse. After school, by the time the child arrived at the pediatrician, however, her condition had deteriorated, and she was given three doses of emergency medication through a nebulizer, an instrument that provides humidity and thus increases absorption in the lungs, along with an oral steroid. The child's breathing still sounded labored though oxygen levels hovered around the low nineties saturation, and the pediatrician recommended admittance to the emergency room.

Over the course of the next five days, the child was admitted to the PICU, given continuous albuterol, injected with steroids, not permitted to eat or drink for over twenty-four hours, and saw at least twenty different caregivers: nurses, doctors, medical students, and specialists. She was visited on average every one to two hours, and each time her oxygen saturation levels, temperature, blood pressure, heart rate, and the sound of her lungs were assessed. Though her oxygen levels rarely fell below ninety, each new caregiver recommended either continuing a course of treatment or making a very minor change during treatment based on his or her assessment and best understanding of the case at that moment.

No single caregiver was exclusively responsible for decision making about the child's course of treatment over the course of five days, though at least on paper, there was a pediatrician overseeing all the residents and nurses who made these decisions individually. But this particular doctor never actually visited the patient. After six days in the hospital, the child was ratcheted down in her medications based on benchmarks specified by the standard of care: ratcheting down the meds according to those assessing the sound of her lungs and recorded oxygen levels, blood pressure, heart rate, and temperature. At night, her oxygen levels would fall, and caregivers would ratchet up the medications and oxygen. During the day, the process would reverse. During the time in the hospital, the child acquired a fungal infection and ate no high-fiber foods, very little protein, and no fresh fruits or vegetables. When released from the hospital, her oxygen saturation levels were hovering around ninety-one (not much different from the day she was admitted), but she "sounded better" according to the most recent caregiver.

This process outlined above is very expensive. When her parents ask the doctors to explain why their child needs to stay in the hospital when it is not clear that her health is all that different from when she arrived, their concerns are not addressed directly. When they ask whether their child might do perfectly well at home with a home nebulizer, a visit or two from a nurse, additional medications as necessary, and regular phone calls from the pediatrician, they are warned that the hospital will call Child Protective Services if the child is removed from the hospital.

There are a number of things wrong with this process. First, it is unclear what the evidential grounds are for the decision process or what assumptions the decision makers are making about the process of asthma recovery. Do they assume (or know) that all or most patients who undergo an asthma episode and whose oxygen levels rise above 90 percent for a period rarely return to the hospital? While the functional relationship between oxygen levels and overall health are relatively well understood, there's no data, as far as could be determined, on the following: How high exactly above 90 percent is sufficient and for how long? How frequently ought the levels be assessed?

More importantly, perhaps, how noisy is this process? That is, how large is the variance around 90 percent among patients who recover? Does an upswing in oxygen saturation levels always or for the most part remain high, or are there occasional downswings? And how frequent, exactly, are these downswings? In fact, high oxygen can be toxic, and "lower levels of oxygen in the blood than are considered normal are not necessarily harmful and may be seen in people who subsequently fully recover, or in healthy people at altitude" (Gilbert-Kawai 2014). Pulse oximeters measure oxygen saturation, which tracks how much hemoglobin is bound to oxygen in the blood. The "oxyhemoglobin dissociation curve" is a sigmoid-shaped curve representing the relation between the partial pressure of oxygen (*x* axis) and the oxygen saturation (*y* axis). Hemoglobin's affinity for oxygen increases as successive molecules of oxygen bind. At 90 percent saturation, the maximum number of hemoglobin molecules are bound. The effectiveness of hemoglobin-oxygen binding can be affected by several factors. For instance, the curve is shifted to the right by an increase in temperature or a decrease in pH. The curve is shifted to the left by the opposite of these conditions. In other words, *a variety of factors could potentially create a great deal of noise in these measures.* It is plausible that oxygen saturation levels vary among individuals and over the course of the day, as do blood pressure and temperature. Oxygen saturation levels typically fall at night, so when they do fall, this is not necessarily reason for alarm; yet their falling is treated the same regardless of time of day.

Second, and central to our discussion, there is the problem of continuity of care; no single caregiver is assessing the child in my example over the course of the decision making process, so no caregiver is familiar with this patient's baseline health, triggers, or typical patterns of exacerbation. Whether this information was adequately communicated among the many different parties is unclear. Third, and relatedly, many of the sequential decisions are made based on listening to labored breathing of the patient, which is a relatively subjective matter. What "sounds bad" to one caregiver may well have sounded "better" to another, and there will likely be some baseline variation in the sound of the lungs among children depending at least in part on how well their asthma is controlled, genetics, and environment, as well as whether their asthma is triggered by a common cold, allergies, or other causes.

Fourth, communication with family members is clearly failing. The patient's parents' well-meaning questions were either inadequately addressed or responded to with inappropriately threatening and paternalistic behavior. By alienating the parents in this encounter, the physician and hospital effectively reinforced the notion that patients and their families are passive recipients of health information rather than active participants in caring for their child's health and equal partners in deciding on her care. Such an experience might be so fear inducing or alienating as to prevent a family's taking a child to the hospital in the future—a possibility that itself carries serious risks.

Finally, even though each decision by each caretaker may well have been "rational" (in some sense, to be discussed below), the overall process of decision making is poor at best. It is not clear that the entire process yielded a lasting overall improvement in the child's health. Of course, estimating these risks and benefits carries a level of uncertainty, so there is some degree of precautionary judgment at stake. Indeed, it is possible that "defensive medicine" is what's driving these lengthy stays in the hospital. Nonetheless, it is unclear whether reasonable, alternative options for systematic patterns of care have been considered. The same end points may well have been achieved at lower cost with either attentive home care or a different decision making strategy. The evidence is simply not there: the assumption that seems to support the current standard of care is that children in such circumstances are at risk of serious harm and that this risk outweighs any risk of harm associated with being in the hospital.

But there has been no controlled trial supporting this assumption. In the United States, at least, no one has yet conducted a trial of hospitalization against monitored home care (drugs, etc., and supervision via either nursing visits or regular phone calls from a pediatrician), which might equally well reduce morbidity and mortality for children with asthma. Surely, this is in part because of concerns about the ethics of such a trial, but these concerns may be based on unfounded assumptions about risk and paternalism about the capacity of patients and families to understand and participate in their own health. It is certainly possible that some proportion of patients currently undergoing days of hospitalization may have done as well, if not better, with carefully monitored home care. Indeed, it is likely that some proportion of patients are being overtreated. Perhaps in part this is because some proportion of patients are underinsured and use the hospital in lieu of regular pediatric care.

It is difficult not to conclude that clinicians—or, perhaps better, the distributed teams of nurses, clinicians, hospital administrators, legal review teams, and hospital oversight committees jointly responsible for such decisions—are not willing to consider a range of options for systematic decision making about treatment of the disease. Indeed, a central challenge in such cases is identifying one central party responsible for making such decisions and thinking through an overall strategy for maximizing effective outcomes. This challenge may well be a significant cause of systematic overtreatment in the United States (Welch, Schwartz, and Woloshin 2012).

In sum, while each decision in this process might be said to be in some sense "good," it is questionable whether the entire system is an effective standard of care. This outcome is in part because of the history of the establishment of this practice. As happens all too often in medicine, standards of care become established as a matter of inertia (i.e., what is done now may be better than what was done before, but it may not be better than all [reasonable] alternatives). Surely, asthma care today is better than it was twenty-five or thirty years ago. But it is still the case that asthma care and, indeed, many medical interventions, are currently driven by "defensive" medical practice rather than well-thought-out action plans for community-based care.

Some standards of care have the inertia they do because they work; but not all do, and it seems that there are very few systematic procedures in place to test alternatives and determine what might be more effective. There is an alternative to this pattern of inertia: consciously rethinking the pattern and process of decision making—and this alternative should be par for the course in medicine. That is, a central part of assessment of effectiveness in medical care should involve the assessment of the effectiveness of decision making, communication, and consideration of relevant alternative plans of organization for decision making. The challenge, of course, is determining how to measure and thus quantify "effective" distributed decision making.

What makes care effective? One aim of this chapter is to argue that the larger picture—the whole *distribution* of care and expense, rather than single decisions in isolation, should figure in our assessments of effective care. I take this to be roughly in keeping with those who argue that we need to think of quality of life as a process (e.g., McClimans and Browne 2012). Once we see that whole distributions of care and expense need to figure into our assessments of effectiveness, then it becomes clear that *a persistent pattern of spreading decision making over time and failing to rethink the larger pattern of care* is a problem of measurement. Indeed, I suggest it is no less a problem, and perhaps a yet more serious one, than those Jacob Stegenga (2015) discusses, namely, the use of poor measuring instruments, the use of misleading analytic measures, and the assumption that measurement in an experimental setting is sufficient to infer properties of a general capacity of effectiveness. For instance, in the case I discuss, while each decision may well be sound overall, a failure to consider alternatives when assessing the effectiveness of a *pattern* of intervention or treatment protocol is as serious a problem as using poor measuring instruments, using misleading analytic measures, and making the assumption of generality. *Good individual decisions—even the most well-tested instrumental interventions—are consistent with bad overall patterns of care.*

EFFECTIVE DECISION MAKING

What does effective decision making look like in clinical contexts? To make this response systematic, it is first necessary to briefly introduce standard utility theory, address some common challenges to standard utility theory, and consider a variety of solutions that have been proposed. To be clear, I am neither endorsing nor challenging standard utility theory as a normative or descriptive theory; my aim is simply to give a semiformal analysis of the decision situation and draw on the literature in decision theory to illustrate the different ways in which medical decision making is constrained by the institutional or structural features of the decision process and the interests of competing stakeholders in the decision process.

Standard expected utility theory provides norms for assessing which choice maximizes expected utility. Put slightly differently, "Expected utility theory provides a way of ranking the acts according to how *choiceworthy* they are: the higher the expected utility, the better it is to choose the act" (Briggs 2015, section 1). The expected utility of an outcome depends on both the likelihood of it coming about, conditional on some action, and the value I assign to that outcome, measured by a number called "utility." Standard utility theory assumes that agents are rational or (what is the same on this view) that they will act in a way that maximizes the chance of their gaining the most utility.

Putting this theory into context, consider a bet, which gives the agent a 50 percent chance of winning $200 and a 50 percent chance of losing $100. Utility maximizers should (all things considered) accept such a bet, provided that their utility function is "linear" (i.e., they want to maximize utility), so they chose options where they stand to gain more than they stand to lose. The average value gained by playing this game is $50, so a "rational" player should accept the bet. Accepting this bet maximizes utility unless one is very risk averse.

The reason this is relevant to our discussion above is as follows. Individual caregivers have good reason to be risk averse. Each decision could potentially incur a great loss *for that caregiver*. Consider a "wrong" decision (e.g., deciding to permit a patient to leave the hospital when in fact he or she is unprepared to go). This decision could result in a relapse and either a return for further treatment or (worse still) death. Such decisions represent a huge disutility, even if a very low probability. Thus, on standard utility theory grounds, any caregiver has exceptionally good reason to err on the side of caution (i.e., keep the patient in the hospital longer until there is less uncertainty).

However, when we compound these decisions, we end up with what, ultimately, is a poor overall outcome for patients. Individually "rational" risk-averse decisions on the part of caregivers may be "high utility" from the perspective of individual caregivers, but systematically lead to bad overall outcomes for patients. For while it is wise from the perspective of any caregiver to err on the side of caution, it may be in patients' best interests to leave the hospital sooner rather than later; it may also be in their best interests to receive far fewer tests and treatments than they are likely to receive (given our current pattern of overtreatment). Considering the perspective that caregivers have and the small but quite serious possibility of medical error, they often choose the "more is better" option (e.g., Welch et al. 2012). Yet overall, the utility of continued tests, treatments, and long stays in the hospital is extremely low.

Agents making the decision may weigh different outcomes differently and be risk averse. From the perspective of classical utility theory, "Good decisions are those that effectively choose means that are available in a situation to achieve as well as possible the individual's goals" (Patel et al. 2002). But what constitutes a good decision in this case depends significantly on factors apart from the individual decision maker's goals and his or her ranking of the value and disvalue of outcomes. Ostensibly, all decision making in medicine ought to be aimed at the *larger goal* of improving the patient's health and well-being. But the fact that the decision is distributed over multiple caregivers means that in each case, the individual making the decision includes some assessment of his or her own risk and precaution, alongside the other costs and benefits.

This picture is complicated still further by intertwined institutional, economic, and psychological factors. First, regarding institutional and economic factors, different hospitals have different hierarchies involved in the decision-making process. In teaching hospitals, for instance, there may be a medical resident in charge of a group of medical students, who in consultation decide on a course of action for each patient. But the decision making process often involves many "micro-decisions," to which the resident and medical students may not be privy. Each assessment of the patient itself involves a decision, and most such "minor" decisions are made not by clinicians but rather by nurses or nursing assistants. Such individuals "lower" down on the hierarchy have more to lose and thus may place greater emphasis on their own individual risk, thus inadvertently practicing defensive medicine.

In nonteaching hospitals, there may be a clinician in charge of a ward and one clinician charged with a given patient for a specific period who supervises a group of nurses for that same period. But shift work such as this means that the individuals making such decisions may change as many as three to five times over the course of twenty-four hours. Although there is a "single" doctor in charge of any patient's case at any given time, that doctor is simply a stand-in for three to five clinicians who may pass through, each arriving at different times during the length of the hospitalization. Nurses also may have longer or shorter exposures to the same patient. In one seventy-two-hour stay in the hospital, a single patient may see as many as thirty-five different doctors, nurses, nursing assistants, and orderlies. Such a situation is further complicated by the fact that everyone making these decisions has a different experience and expertise, so each decision is subject not only to an individual's varying calculations of risk but also informed by potentially very different background beliefs and knowledge bases.

Second, with respect to psychological factors, in contexts where decisions are made by teams, there are several dynamics involved. Even if a single expert individual is ultimately responsible for deciding on a course of action, others are critically involved in the process and may sway an individual, perhaps against his or her better judgment. Such decisions are embedded in the broader context of the hospital. This context includes standards and risk tolerance; it also includes decision-action cycles, which may be variously affected by monitoring and feedback not only from other medical experts but also from the upper administration, who have different goals than clinical staff. Moreover, cognitive features of groups are quite different from those of individuals, and modeling the dynamics of these decision processes is complex.

Also, with respect to psychological factors, the level of individual decision making, availability biases may make certain decision makers particularly averse to risky decisions. The availability bias is the tendency to assess the frequency, probability, or likely causes of an event by the degree to which the instances or occurrences of the event are readily available in memory. For example, the loss of a patient due to asthma will be a vivid, distinct, easily imagined, and specific event that will strongly influence a decision maker. It will be "more available" and emotionally salient to a physician or nurse than a proper consideration of base rates or actual likelihood of such an outcome. Moreover, "group think" can create unhealthy dynamics; a group may sway a clinician who otherwise is making a relatively sound overall decision in favor of a less sound decision.

On top of all these considerations, decisions in real-world clinical settings often must be made under constraints, such as stress, time pressure, and limited resources. In emergency and triage situations or intensive care medicine, decisions are often made with very limited time and information available. Taking all such matters into consideration, we arrive at a pattern of decision making that—by and large—tends to be highly risk averse and is complicated not only by individual biases but also by influences of group dynamics and the larger context of hospital administrators, insurers, and patients and families as well as the larger social context (e.g., a litigious environment). This situation is complicated further by the fact that real-life decision makers tend to be risk averse, to the extent that they violate some of the fundamental principles of rational decision theory (e.g., Allais 1979).

To be sure, how to assess the overall utility in such a case depends on a variety of factors. Nonetheless, what kinds of decisions are more likely to result in a desirable overall outcome, all things considered? The problem with the narrow view of utility as a matter of what is of value for a decision maker at a particular time is that it does not reflect the fact that decision making in medicine is distributed. Thus, utility is a "one-off" matter of what risk/benefit calculation we should make for any administrative agent who is making the decision. When assessing the utility of such decisions, agents are sensitive to their interests and preferences—the risk to their career of a bad outcome, the cost of communicating and enforcing the decision (e.g., populating the appropriate database), ensuring that those carrying out this process do in fact follow through—as well, of course, as the patient's health and well-being, administrative costs, insurance billing procedures, and so on. But each decision taken individually may well maximize utility, all things considered, yet fail to improve patient health overall.

My point is this: effective care is not strictly a one-dimensional problem. "Effective" health care can refer to narrow "clinical" efficacy—as a drug may be effective at preventing or curing a disease—but it can also refer to effective allocation; cost effectiveness; effectiveness in the management of financial risk; and effective "timing" or instrumental use of medicines, equipment, products, or services. In addition, effectiveness in considering decision making patterns, the widest possible alternatives outside the standard of care, communication between patients and clinicians, and attention to the dynamic and distributed nature of care and patient decision making are all of relevance to the effectiveness of care.

I have suggested that this collectively risk-averse decision process is less than optimal. It results in unnecessary lengthy stays and overtreatment for many patients, at great cost to both families and hospitals. Many patients are unlikely to benefit from close monitoring and extensive treatments and tests, and they may even be harmed by these interventions.

THE GROUP AND THE INDIVIDUAL

The family of problems associated with the decision making problem I have been discussing are not independent of a second, far more familiar problem. This is the very problem that led to personalized, or "precision," medicine. Namely, even if we identify all potentially relevant reference classes to which a given patient belongs and have detailed clinical and epidemiological data on the risk and benefits associated with a given course of action for individuals belonging to that reference class, it is possible that the chosen course of action is inappropriate (i.e., harmful for that patient). Why? Because each patient is in some sense unique. This point about external validity has been made forcefully by others (e.g., Cartwright 2007; Cartwright and Munro 2010).

This issue is made vivid in the case of asthma patients because the risk factors associated with asthma and the nature of the condition vary significantly. How and why a patient responds to different treatments may well depend not only on relatively rare or idiosyncratic factors, such as allergic reactions to latex, but also could well depend on relatively common but unknown factors. Why and how frequently a patient benefits (or not) from hospitalization may well be a relatively idiosyncratic matter, having to do with not only the genetics or basic physiological response of the patient but also with everything from home environment to stressors associated with asthma, which vary significantly from patient to patient.

The parallels between rational decision making and questions of external validity are quite striking. In the section above, we were considering how each decision may be rational, but overall, the pattern of decisions could lead to poor outcomes. Here, we are considering that what may be effective for most patients with a particular condition, may be less so for a specific patient. In both cases, what may seem "rational" on a very narrow interpretation of rationality, or "effective" on a very narrow interpretation of effectiveness, is in fact less than optimal, all things considered. In both cases, attention to the wider view (i.e., the long-term process of patient care or the heterogeneity of patient populations) can lead one to rethink what counts as rational and what counts as effective.

SOLUTIONS

Several solutions have been proposed in the medical decision making literature, none of which, in my view, seem sufficiently complete or satisfactory. The reason is that they focus on the problem of communication between clinicians and their coworkers, for instance, in the NICU. Several complicating factors arise in such cases: lack of transparency or efficacy in communication, challenges in arriving at collaborative decisions, stress, time limitations, response variability, and challenges in overcoming differences in knowledge and expertise. In my view, none of these studies get at the fundamental problem of overall patterns of care; distributed decision making; and risk aversion, or "passing the buck." No caregiver wishes to be responsible for making a decision that results in serious medical error. If these studies do not address this issue, how might we?

My suggestion is that we measure and assess multiple "dimensions" of effectiveness when assessing quality of care. Were we to consider these dimensions, we might change our patterns of assessment of quality of care. Unfortunately, the problem of measuring "effectiveness" in health care is often framed as a narrow empirical problem (i.e., What are the best methods of evidence gathering for and against various treatments and preventive measures?). However, effectiveness is not strictly a one-dimensional problem. As mentioned above, narrow "clinical" efficacy, even effective allocation, cost effectiveness, and effectiveness in the management of financial risk, does not capture the fact that considering decision making patterns is essential to benefiting the patient. Reconsidering how patient decision making is distributed as well as willingness to reengineer the process to prevent the kind of systematic risk aversion, are essential to better care. More generally, caregivers must consider alternatives to the standard of care, transparency, and communication, where patients are better able to understand and decide about their treatment, with better education and more transparent and effective communication about the known risks and benefits of intervention. Part of the argument here is that including patients and families in their care—empowering them with the knowledge of how to be healthier as a matter of baseline health—is as important if not more important a dimension of "effectiveness" of care than the reduction of an end point for any particular intervention. Community-based care may both shorten the length and number of hospitalizations for everything from childhood asthma to heart disease. All these dimensions are, in my view, components of effective medical care.

REFERENCES

Allais, Maurice. 1979. "The Foundations of a Positive Theory of Choice Involving Risk and a Criticism of the Postulates and Axioms of the American School." In *Expected Utility Hypothesis and the Allais Paradox*, edited by Maurice Allais and Ole Hagen, 27–145. Dordretch, Netherlands: Reidel.

Briggs, Rachel. 2017. "Normative Theories of Rational Choice: Expected Utility." In *The Standford Encyclopedia of Philosophy* (Spring 2017 Edition), edited by Edward N. Zalta. https://plato.stanford.edu/archives/spr 2017/entries/rationality-normative-utility/.

Cartwright, Nancy. 2007. "Are RCTs the Gold Standard?" *BioSocieties* 2: 11–20.

Cartwright, Nancy, and Eileen Munro. 2010. "The Limitations of Randomized Controlled Trials in Predicting Effectiveness." *Journal of Evaluation in Clinical Practice* 16: 260–66.

Gilbert-Kawai, E.T., K. Mitchell, D. Martin, J. Carlisle, and M. P. W. Grocott. 2014. "Low Blood Oxygen Levels versus Normal Blood Oxygen Levels in Ventilated Severely Ill People." Cochrane Review. www.cochrane.org/CD009931/ANAESTH_low-blood-oxygen-levels-versus-normal-blood-oxygen-levels-in-ventilated-severely-ill-people.

Kohn, Linda T., Janet M. Corrigan, and Molla S. Donaldson. 2000. *To Err Is Human: Building a Safer Health System*. Institute of Medicine. Washington, DC: National Academies Press.

Kukla, Rebecca. 2007. "How Do Patients Know?" *Hastings Center Report* 37: 27–35.

Lemieux-Charles, Louise, and Wendy L. McGuire. 2006. "What Do We Know about Health Care Team Effectiveness? A Review of the Literature." *Medical Care Research and Review* 63: 263–300.

Makary, Martin A., and Michael Daniel. 2016. "Medical Error—the Third Leading Cause of Death in the US." *British Medical Journal* 353: i2139.

McClimans, Leah, and John P. Browne. 2012. "Quality of Life Is a Process Not an Outcome." *Theoretical Medicine Bioethics* 33: 279–92.

Patel, Vimla L., David R. Kaufman, and Jose F. Arocha. 2002. "Emerging Paradigms of Cognition in Medical Decision-Making." *Journal of Biomedical Informatics* 35: 52–75.

Patel, Vimla, Jiajie Zhang, Nicole A. Yoskowitz, Robert Green, and Osman R. Sayan. 2008. "Translational Cognition for Decision Support in Critical Care Environments: A Review." *Journal of Biomedical Informatics* 41: 413–31.

Stegenga, Jacob. 2015. "Measuring Effectiveness." *Studies in History and Philosophy of Biological and Biomedical Sciences* 54: 62–71.

Welch, H. Gilbert, Lisa Schwartz, and Steve Woloshin. 2012. *Overdiagnosed: Making People Sick in Pursuit of Health*. Boston: Beacon Press.

Index

About the Contributors

Gabriele Badano is a postdoctoral researcher at the Centre for Research in the Arts, Social Sciences and Humanities (CRASSH) and a junior research fellow at Girton College, University of Cambridge. In addition to research interests in political liberalism and other debates in political philosophy, he has worked extensively on the ethics of the allocation of health resources. His research has been published in venues including *Ethical Perspectives*, *Critical Review of International Social and Political Philosophy*, and *Social Theory and Practice*.

Norman M. Bradburn is the Tiffany and Margaret Blake Distinguished Service Professor Emeritus at the University of Chicago. A social psychologist, Bradburn has been at the forefront in developing theory and practice in the field of sample survey research. He has focused on psychological well-being and assessing quality of life, particularly using large-scale sample surveys, nonsampling errors in sample surveys, and research on cognitive processes in responses to sample surveys. His book *Thinking About Answers: The Application of Cognitive Process to Survey Methodology* (coauthored with Seymour Sudman and Norbert Schwarz, 1996) follows three other publications on the methodology of designing and constructing questionnaires: *Polls and Surveys: Understanding What They Tell Us* (with Seymour Sudman, 1988), *Asking Questions: A Practical Guide to Questionnaire Construction* (with Seymour Sudman, 1982; second edition with Brian Wansink, 2004), and *Improving Interviewing Method and Questionnaire Design* (1979).

Alex Broadbent is professor of philosophy and executive dean of the faculty of humanities at the University of Johannesburg. Previously, he held various research, teaching, and visiting positions at Cambridge, Vienna, Athens, and

Harvard before joining the University of Johannesburg in 2011. Alex is a philosopher of science with interests in philosophy of epidemiology, philosophy of medicine, and philosophy of law, connected by the philosophical themes of causation, explanation, and prediction. He is committed to finding philosophical problems in practical contexts and to contributing something useful concerning them. He holds a P rating from the National Research Foundation of South Africa (2013–2018) and is a member of the South African Young Academy of Sciences. He has published several articles in top-ranked international journals across three disciplines (philosophy, epidemiology, and law). His first book, *Philosophy of Epidemiology*, was published in 2013 and has been translated into Korean. His second book, *Philosophy for Graduate Students: Metaphysics and Epistemology*, was published in February 2016. He is currently working on his third book, *Philosophy of Medicine*.

Nancy Cartwright is a methodologist and philosopher of the natural and social sciences, with special focus on causation, evidence, and modeling. Her recent work has been on how to make the most of evidence in evidence-based policy. She is a professor of philosophy and codirector of the Centre for Humanities Engaging Science and Society (CHESS) at Durham University in the United Kingdom and distinguished professor of philosophy at the University of California San Diego, having worked previously at Stanford University and the London School of Economics. She is a MacArthur Fellow; a fellow of the British Academy, the American Philosophical Society, and the Academy of Social Sciences; and a member of Leopoldina (the German National Academy of Natural Science) and the American Academy of Arts and Sciences.

Laura Cupples was awarded a BA in physics from Davidson College in 2002 and an MA in philosophy from the University of South Carolina in 2013. She is currently a PhD candidate in philosophy at the University of South Carolina, where she holds a Social Advocacy and Ethical Life Presidential Doctoral Teaching Fellowship. She specializes in philosophy of measurement in the human sciences and philosophy of science in practice more generally. Her dissertation work focuses on the epistemology of health-related quality of life measurement, including the epistemic roles of models and construct theories in psychometrics. Other academic interests include history of science and technology, science and values, social and feminist epistemology, and applied ethics.

Michael Dickson is professor of philosophy in the College of Arts and Sciences at the University of South Carolina, where he has been since 2004. Prior to moving to South Carolina, he was Ruth N. Halls Professor of History and Philosophy of Science at Indiana University. He works primarily in philosophy of science, especially the foundations of quantum theory, but has also done work in general philosophy of science and the relationship between science and metaphysics. More recently, his work has been in game theory and related fields.

Eivind Engebretsen is professor of medical epistemology and research director (as the first with a human science background) at the faculty of medicine, University of Oslo, Norway. His current research is concerned with the discourse of "knowledge translation" within medicine and its different genealogies and how it might be expanded by drawing on theories of translation from linguistics, philosophy, and anthropology. He leads the research group KNOWIT—Knowledge in Translation.

Jonathan Fuller is a philosopher of medicine in Toronto. He completed a PhD in the philosophy of medicine and a research fellowship in health professions education at the University of Toronto in 2016. He is currently studying medicine and is a curriculum lead for the theme "Foundations of Knowledge" in Toronto's MD program.

Daniel M. Hausman is the Herbert A. Simon and Hilldale Professor of Philosophy at the University of Wisconsin-Madison. A founding editor of the journal *Economics and Philosophy*, his research has centered on epistemological, metaphysical, and ethical issues lying at the boundaries between economics and philosophy. His most recent books are *Preference, Value, Choice and Welfare* (2012); *Valuing Health: Well-Being, Freedom, and Suffering* (2015); and *Economic Analysis, Moral Philosophy, and Public Policy*, third edition (coauthored with Michael McPherson and Debra Satz, 2017).

Kristin Heggen is professor in health sciences, University of Oslo. She is currently serving as dean of education at the faculty of medicine, University of Oslo. Heggen has expertise in the humanities, the social sciences, and educational research. Her research interests comprise ethical issues and power dynamics in health care and issues concerning education and knowledge transfer between academic and clinical settings.

Stephen John is the Hatton Senior Lecturer in Philosophy of Public Health at the department of history and philosophy of science, University of Cambridge. His research focuses on questions at the intersection of political philosophy, philosophy of science, and public health policy. He is currently co-PI of a project on the "Limits of the Numerical," which considers the causes and effects of quantification in health care.

Trenholme Junghans is a cultural anthropologist and research associate at the University of Cambridge, where she is affiliated with the "Limits of the Numerical" project at the Centre for Research in the Arts, Social Sciences and Humanities (CRASSH). She received her PhD from the University of St. Andrews, where she was jointly supervised in social anthropology and management. She is broadly interested in understanding the world-making propensities of processes of abstraction and classification, of quantification and commensuration. Trenholme's interests and experience are broadly interdisciplinary in scope, spanning contemporary social theory, science and technology studies, semiotic and linguistic anthropology, and political economy, and she has conducted extensive anthropological research in Hungary.

Aaron Kenna earned his MS in philosophy of science from the University of Utah and currently is a PhD candidate in the history and philosophy of science at the University of Toronto's Institute for the History and Philosophy of Science and Technology. His interests lie principally in the philosophy of science, philosophy of medicine, and philosophy of statistics, with an emphasis on the use of statistical methodology in scientific reasoning. Recently, he was awarded a Junior Philosophy of Science Fellowship from Ludwig Maximilian University of Munich's Center for Mathematical Philosophy to study the role of Bayesian epistemology in quantifying deep uncertainties in climate change science.

Leah McClimans is associate professor in the philosophy department at the University of South Carolina and women's and gender studies program. Her research interests lie at the intersection of philosophy of medicine and medical ethics. She has authored numerous articles on measurement in quality-of-life research, clinical ethics, and the entanglement of ethics and evidence in questions about place of birth and elective reductions. Before coming to the University of South Carolina, she held a postdoctoral fellowship at the University of Toronto's Joint Centre for Bioethics (2006–2007). She also held an Ethox Research Fellowship (2009–2010) at the University of Warwick Medical School and has been awarded a Marie Curie ASSISTID fellowship to address the ethical and epistemological questions regarding the evidence base for assistive technologies (2017–2019). In 2015, she was named a Distinguished Faculty Member in the College of Arts and Sciences at the University of South Carolina, and in 2016 she was awarded the Provost's Mungo Undergraduate Teaching Award.

Zinhle Mncube is a lecturer in the department of philosophy at the University of Johannesburg in South Africa. Her research interests lie broadly in philosophy of medicine, philosophy of biology, and philosophy of race. She has published an article on the biological basis of race titled "Are Human Races Cladistic Subspecies?" and her master's dissertation was on how we might causally construe heritability estimates. Zinhle lectures undergraduate courses in philosophy of race. She is also an Iris Marion Young Scholar and a Cornelius Golightly Fellow.

Anya Plutynski is an associate professor of philosophy at Washington University in St. Louis, where she has been since 2013. She was also a professor at the University of Utah, from 2001–2013. Her areas of expertise are history and philosophy of biology and medicine. Her books include the Blackwell's *Companion to Philosophy of Biology* (coedited with Sahotra Sarkar) and the *Routledge Handbook of Philosophy of Biodiversity* (coedited with Justin Garson and Sahotra Sarkar). She is currently completing a book forthcoming, *Explaining Cancer: Philosophical Issues in Cancer Classification, Causation and Evidence*. She is also associate editor at the *British Journal for Philosophy of Science*.

Benjamin Smart is a senior lecturer in philosophy at the University of Johannesburg. Prior to relocating to South Africa in 2015, he lectured at the University of Birmingham for two years. He was awarded his PhD from the University of Nottingham in 2012. Smart writes primarily on the philosophy of medicine and metaphysics. In 2016, he published a monograph titled *Concepts and Causes in the Philosophy of Disease*.

Jacob Stegenga is a lecturer in the department of history and philosophy of science at the University of Cambridge. He held the prestigious Banting Postdoctoral Fellowship at the University of Toronto and has been a fellow at the Institute of Advanced Study at Durham University. His research focuses on philosophy of science, including methodological problems of medical research, conceptual questions in evolutionary biology, and fundamental topics in reasoning and rationality. His present work is culminating in a book titled *Medical Nihilism*, in which he argues that if we attend to the extent of bias in medical research, the thin theoretical basis of many interventions, and the malleability of empirical methods in medicine, and if we employ our best inductive framework, then our confidence in medical interventions ought to be low.

Ross Upshur (MD) is currently the scientific director for the Bridgepoint Collaboratory for Research and Innovation and assistant director of the Lunenfeld Tanenbaum Research Institute, Sinai Health System. At the University of Toronto, he is a professor in the Dalla Lana School of Public Health (and head of the division of clinical public health) and in the department of family and community medicine. He is an affiliate member of the Institute for the History and Philosophy of Science and Technology and adjunct senior scientist at the Institute for Clinical Evaluative Sciences. In 2015, Dr. Upshur was named one of the Top 20 Pioneers in Family Medicine Research and Family Medicine Researcher of the Year, by the College of Family Physicians of Canada. Dr. Upshur is the former director of the University of Toronto Joint Centre for Bioethics (2006–2011) and was Canada research chair in primary care research (2005–2015). He is a member of the Royal College of Physicians and Surgeons of Canada and the College of Family Physicians of Canada.

www.ingramcontent.com/pod-product-compliance
Ingram Content Group UK Ltd.
Pitfield, Milton Keynes, MK11 3LW, UK
UKHW041842031026
15587UKWH00002B/951

9 781783 488476